T0178397

An Introduction to Inertial Confinement Fusion

Series in Plasma Physics

Series Editor:

Steve Cowley,
Imperial College, UK, and University of Los Angeles, USA

Other books in the series:

Aspects of Anomalous Transport in Plasmas
R Balescu

Non-Equilibrium Air Plasmas at Atmospheric Pressure
K H Becker, R J Barker and K H Schoenbach (Eds)

**Magnetohydrodynamic Waves in Geospace: The Theory of ULF Waves and their
Interaction with Energetic Particles in the Solar-Terrestrial Environment**
A D M Walker

Plasma Physics via Computer Simulation (paperback edition)
C K Birdsall, A B Langdon

Plasma Waves, Second Edition
D G Swanson

Microscopic Dynamics of Plasmas and Chaos
Y Elskens and D Escande

Plasma and Fluid Turbulence: Theory and Modelling
A Yoshizawa, S-I Itoh and K Itoh

The Interaction of High-Power Lasers with Plasmas
S Eliezer

Introduction to Dusty Plasma Physics
P K Shukla and A A Mamun

The Theory of Photon Acceleration
J T Mendonça

Laser Aided Diagnostics of Plasmas and Gases
K Muraoka and M Maeda

Reaction-Diffusion Problems in the Physics of Hot Plasmas
H Wilhelmsson and E Lazzaro

The Plasma Boundary of Magnetic Fusion Devices
P C Strangeby

Non-Linear Instabilities in Plasmas and Hydrodynamics
S S Moiseev, V N Oraevsky and V G Pungin

Collective Modes in Inhomogeneous Plasmas
J Weiland

Series in Plasma Physics

An Introduction to Inertial Confinement Fusion

S Pfalzner
University of Cologne, Germany

CRC Press
Taylor & Francis Group
Boca Raton London New York

CRC Press is an imprint of the
Taylor & Francis Group, an **informa** business

CRC Press
Taylor & Francis Group
6000 Broken Sound Parkway NW, Suite 300
Boca Raton, FL 33487-2742

First issued in paperback 2019

© 2006 by Taylor & Francis Group, LLC
CRC Press is an imprint of Taylor & Francis Group, an Informa business

No claim to original U.S. Government works

ISBN-13: 978-0-7503-0701-7 (hbk)
ISBN-13: 978-0-367-39109-6 (pbk)

Library of Congress Card Number 2005046678

Library of Congress Cataloging-in-Publication Data

Pfalzner, Susanne.
An introduction to inertial confinement fusion / Susanne Pfalzner.
p. cm. -- (Series in plasma physics)
Includes bibliographical references and index.
ISBN 0-7503-0701-3 (acid-free paper)
1. Pellet fusion. I. Title. II. Series.

QC791.775.P44P43 2006
539.7'64--dc22   2005046678

**Visit the Taylor & Francis Web site at
http://www.taylorandfrancis.com**

**and the CRC Press Web site at
http://www.crcpress.com**

Dedication

To my husband, Paul,
and my daughters, Theresa and Claire

I am one of those who believe like Nobel that mankind will derive more good than harm from the new discoveries.

Pierre Curie

Preface

The idea for this book arose from talking to some colleagues about how confusing it was at one's first inertial confinement fusion (ICF) conference to find all these different subjects presented with cryptic names such as SBS, random phase plates, and mode coupling, but one was not able to gain an overall picture of ICF. It turned out that for each of us, it had been a rather lengthy and painful process to create this picture for oneself. Although there do exist a number of very good books about the state of the art of ICF, they are mainly aimed at specialists. The purpose of this book is to assist future newcomers to this field by giving a general overview of the processes involved in inertial confinement fusion on a more accessible level.

Any work trying to encapsulate a rapidly evolving field faces the issue of timeliness. However, the timescale at which material becomes outdated depends on the type of information conveyed. Although there have been many exciting and important new developments in ICF during the last 10 years, there is a core of "conventional understanding" that has remained substantially intact. In other words, numbers change much more rapidly than the ideas behind them. It is these ideas that form the basis of the present text.

A further difficulty that arises with ICF is to present it in a logical order. The plan chosen in this book is to first give an overview of the subject and then follow the ICF process from the driver technology to the burn physics in its chronological order, explaining the physical concepts and obstacles encountered on the way. This is rounded off with a look into the future, (i.e., possible reactor designs and alternative routes).

The book is nominally aimed at physics graduate students. The prospective reader is assumed to have a solid background in physics at the undergraduate level. I do not assume a similar training in plasma physics, but provide a short overview of the relevant plasma phenomena in Chapter 3. The complete text contains somewhat more material than can be covered in a single semester and is to a large extent based on a lecture course given in the winter semester 2004/2005 entitled "Energy like in the sun? An introduction to inertial confinement fusion" (translation from the German)

Over the years, I have received generous help and advice from many individuals, and it is a pleasure to acknowledge them. I am also indebted to Paul Gibbon for his tireless efforts in reading the entire manuscript and offering invaluable commentary on both the scientific content and the manner of its presentation. I am grateful to the students of Cologne University for their comments and corrections. Many thanks as well to A. R. Bell who communicated the initial synopsis to Institute of Physics. I want to thank S. Atzeni, S. Eliezer, J. Jacobs, R. L. McCrory, and S. Nakai for providing me access to figures.

Because my own research has recently moved toward astrophysical applications, some inaccuracies might have crept in for which I am, of course, solely responsible. I trust these will be at worst of detail and not of principle.

Finally, it is a pleasure to thank John Navas of Institute of Physics Publishing, who gave much friendly and helpful advice in seeing this work through to fruition.

Susanne Pfalzner

Contents

Chapter 1

Fundamentals of Inertial Confinement Fusion

It has long been a dream of mankind to produce energy just the way the sun does. Since the early 20th century, we have known that the source of energy for the sun — as for other stars — is a process called nuclear fusion, yet civil research activities in this field did not start until the 1950s. Nowadays many countries support fusion research in the quest for a new resource for the production of electricity. Performing such research is increasingly important as the energy problem becomes a more and more pressing issue (for a brief summary, see Appendix A.1).

Fusion could be one of the solutions to the energy problem, especially because it offers ecological and safety advantages compared with burning coal and oil or nuclear fission power plants. In addition, fusion has the very attractive feature that the fusion fuel can be extracted from seawater, making it directly available to most countries in the world.

Although significant progress in fusion science and technology has been made, to date no practical fusion reactor is running. As a first step to understanding inertial confinement fusion, we will address the question of how the sun produces the energy that forms the basis of all life on Earth.

1.1 What Happens in the Sun?

To answer this question, we have to go back to the basics of nuclear physics. The key to nuclear fusion reactions and possible energy release is the binding energy in the nuclei. Einstein showed that mass and energy are connected via the relation

$$\Delta E = \Delta m c^2. \tag{1.1}$$

1

$$A^{1/3} \qquad\qquad A \qquad A^{3/4}$$
$$\tag{1.2}$$

where m_n and m_p are the neutron and the proton mass, a_v, a_s, a_c, a_a, and a_p, are constants found by fitting experimental binding energies and δ is an odd-even term (see Appendix B.4). The binding energy B of the nucleus is then the difference of the mass of the components (protons and neutrons) and the complete nucleus; that is

$$B = Zm_p + Nm_n - M, \tag{1.3}$$

giving the mass in energy units ($c^2 = 1$). This is the energy needed to separate all nucleons to distances where they no longer interact. Using Eqs. 1.2 and 1.3, one obtains for the binding energy per nucleon

$$B/A = a_v - a_s A^{-1/3} - a_c \frac{Z(Z-1)}{A^{4/3}} - a_a \frac{(N-Z)^2}{A^2} - \frac{a_p \delta}{A^{7/4}}. \tag{1.4}$$

Fig. 1.1 shows the binding energy per nucleon as a function of A. This relatively smooth function shows a broad maximum in the region for nuclei near iron, which are the most stable nuclei. For nuclei much lighter or much heavier than iron, the binding energy per nucleon is considerably smaller. This difference is the basis for fusion and fission processes. The basis of nuclear fusion is that if two very light nuclei fuse, they form a nucleus with a higher binding energy (or lower mass), thus releasing energy according to Einstein's famous formula (Eq. 1.1). Energy is also released when a heavy nucleus splits into two smaller fragments — fission.

In principle there are many energy-releasing fusion processes between different low mass elements possible. However, the problem in igniting such a fusion reaction is that the light nuclei are positively charged and strongly repel each other, so that under normal conditions the distance between nuclei is so large that a nuclear reaction is quite unlikely. So how can this lead to such powerful energy production in the sun anyway? Because of high temperature ($\sim 10^6$ K) and pressure in the center of the sun, the large number of particles, and the relatively long time span available, the cross section for such reactions is still large enough to maintain the huge energy releases characteristic of the sun.

In the sun the energy is mainly obtained from a cycle of proton-proton reactions. They can be summarized as

$$
\begin{aligned}
p + p &\longrightarrow D + e^+ + 2\nu_e & 0.42 \text{ MeV} \\
D + p &\longrightarrow {}^3\text{He} + \gamma & 5.5 \quad \text{MeV} \\
{}^3\text{He} + {}^3\text{He}^{++} &\longrightarrow {}^4\text{He} + 2p & 12.86 \text{ McV}.
\end{aligned}
\tag{1.5}
$$

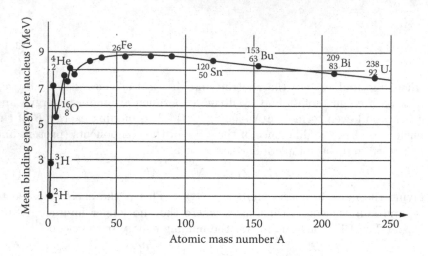

Figure 1.1. Mean binding energy per nucleon of stable nuclei as a function of mass number.

Altogether, the chain reaction leads to the transformation of four protons into ^{4}He summarized as

$$4p \longrightarrow \quad ^4\text{He} + 2e^+ + 2\nu_e + 24.7 \text{ MeV}. \tag{1.6}$$

To a lesser degree, other fusion processes using different reaction cycles leading to the formation of helium take place at the same time (for a more detailed description see (Hodgson *et al.*, 1997; Bahcall and Waxman, 2003).

After a long complex journey through the sun, the energy carried by the gamma rays is eventually transformed into visible light, which it radiates into the surrounding universe. It is this radiation that makes life on Earth possible.

More massive or older stars can use different fusion reactions to produce energy. The above hydrogen-burning process obviously ends when most of the star's hydrogen inventory is burned up. If the star has enough mass, the next type of burning process can start, triggered by the star contracting by gravitational collapse, pushing the temperature up to 10^8 K, making helium burning possible. The fusion of ^{4}He leads to ^{8}Be and eventually ^{12}C. When the helium is exhausted, provided the star mass is sufficiently large, gravitational collapse may again increase the temperature (to about 2×10^9 K) and carbon and oxygen may start to burn, producing neon, magnesium, silicon, phosphorous, and sulfur. At temperatures ranging from 2–5 $\times$ 10^9 K, heavier nuclei up to $A \sim 56$ are produced by

of the star depends very much on its mass and can be, for example, a supernova explosion, a neutron star or even a black hole. For the details on the stellar development, the reader is referred to the relevant astrophysical text books.

1.2 Can One Produce Energy on Earth Like in the Sun?

The process of the energy production of the sun is more — or less — completely understood. So why don't we do it just the same way? The problem is that here on Earth, there is not the space and time the sun has for producing energy. Producing energy on a large scale requires that a huge number of reactions take place together. Coulombic repulsion hinders nuclei from fusing, but one can overcome this by giving the nuclei a high initial kinetic energy, which can be achieved by heating the material to very high temperatures. This approach to fusion is known as "thermonuclear fusion". Energy can either be released in a controlled fashion with a fusion reactor or in an uncontrolled manner using a thermonuclear bomb. From the latter (i.e., hydrogen bombs) we know that thermonuclear fusion is possible, the problem is to do it in a controlled and meaningful way.

Because of the high temperatures and densities required for fusion, the fuel has to be in the plasma state — a hot, highly ionized, electrically conducting gas. If the temperatures are high enough, the thermal velocities of the nuclei become very high. Only then do they have a chance to approach each other close enough so that the Coulombic repulsion can be overcome and the short-range attractive nuclear forces (effective over distances $\sim 10^{-15}$m) can come into play. At this point the nuclei can fuse and free the enormous power encapsulated as shown in Fig. 1.1.

However, under these conditions matter tends to fly apart very quickly unless constrained in some way. In the sun this is done by gravitational forces. As gravity is not a terrestrial option, the central problem is to devise other means of confinement so that conditions of high temperature and density are maintained simultaneously for a sufficiently long time. However, the higher the temperature and density, the more difficult it becomes to confine the plasma. It therefore makes sense to look for a situation where the requirement for confinement — and correspondingly temperature and density — is as low as possible. This is directly linked to the question of which fusion reaction is most readily achieved under these conditions.

Even if the energy of the particles is slightly less than that required to

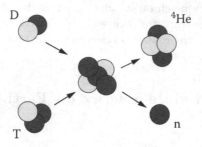

Figure 1.2. Schematic picture of DT-reaction.

overcome the Coulomb barrier, fusion processes can still occur via tunneling. However, the closer the particle energy is to overcoming the Coulomb barrier, the more tunneling processes are likely to happen. To fuse enough particles, the thermal energy of the nuclei should not be too much smaller than their repulsive Coulomb barrier B, which is

$$B \sim 1.44 \frac{q_1 q_2}{r_1 + r_2} \,\text{MeV}$$

where $q_{1,2}$ and $r_{1,2}$ are the charges and radii of the particles in units of the elementary charge and the radii in fm, respectively. For a more detailed description of nuclear processes, see, for example Hodgson *et al.* (1997).

Above we saw that most energy is released if two very light nuclei fuse, for example hydrogen. The Coulomb barrier of two hydrogen nuclei is about 700 keV. Heating the gas to equivalent temperatures would mean $2B/3k_B \simeq 3.6 \times 10^9$ K, which is not a realistic prospect at the moment. Luckily the nuclei of the heavier hydrogen isotopes have smaller Coulomb barriers to overcome, though the energy yield is lower. The fusion reaction of deuterium and tritium turns out to be the easiest approach to fusion because of a relatively large cross section and a very high mass defect (Post, 1990). When these two nuclei (of hydrogen isotopes) fuse, an intermediate nucleus consisting of two protons and three neutrons is formed in the process. This nucleus splits immediately into a neutron of 14.1 MeV energy and an α-particle of 3.5 MeV,

$$^2_1\text{D} + {}^3_1\text{T} \rightarrow {}^4_2\text{He} + {}^1_0 n + 17.6 \,\text{MeV}. \tag{1.7}$$

This fusion reaction has the advantage that the fuel resources are virtually

$$
\begin{aligned}
\text{D} + \text{D} \quad &\longrightarrow \text{T}^4 \ (1.01 \ \text{MeV}) + \text{p} \ (3.03 \ \text{MeV}) \\
&\longrightarrow \text{He}^3 (0.82 \ \text{MeV}) + \text{n} \ (2.45 \ \text{MeV}) \\
\text{D} + \text{He}^3 \quad &\longrightarrow \text{He}^4 \ (3.67 \ \text{MeV}) + \text{p} \ (14.67 \ \text{MeV}) \\
\text{T} + \text{T} \quad &\longrightarrow \text{He}^4 + \text{n} + \text{n} \ (11.32 \ \text{MeV}) \\
\text{He}^3 + \text{T} \quad &\longrightarrow \text{He}^4 + \text{p} + \text{n} \ (12.1 \ \text{MeV}) \\
&\longrightarrow \text{He}^4 \ (4.8 \ \text{MeV}) + \text{D} \ (9.5 \ \text{MeV}) \\
&\longrightarrow \text{He}^5 \ (2.4 \ \text{MeV}) + \text{p} \ (11.9 \ \text{MeV}) \\
\text{p} + \text{Li}^6 \quad &\longrightarrow \text{He}^4 \ (1.7 \ \text{MeV}) + \text{He}^3 \ (2.3 \ \text{MeV}) \\
\text{p} + \text{Li}^7 \quad &\longrightarrow 2\text{He}^4 \ (22.4 \ \text{MeV}) \\
\text{D} + \text{Li}^6 \quad &\longrightarrow 2\text{He}^4 \ (22.4 \ \text{MeV}) \\
\text{p} + \text{B}^{11} \quad &\longrightarrow 3\text{He}^4 \ (8.682 \ \text{MeV}) \\
\text{n} + \text{Li}^6 \quad &\longrightarrow \text{He}^4 \ (2.1 \ \text{MeV}) + \text{T} \ (2.7 \ \text{MeV})
\end{aligned}
$$

unlimited. Deuterium can be produced from sea water, whereas tritium can be generated by lithium reacting with neutrons directly in the reactor. Lithium is relatively abundant on Earth and resources are likely to be sufficient for several 10^4 years. However, using this reaction in a reactor has two disadvantages: tritium is a radioactive gas and lithium a highly poisonous substance. This means that for a reactor design safety is still a major issue. We will address this problem later in Chapter 9. Nevertheless, compared with fission reactors, these problems are relatively minor, given that the half-life of tritium is 12.5 years compared with 2.4×10^7 years for uranium 236, 7.13×10^8 years for uranium 235, 4.5×10^9 years for uranium 238, 24000 years for plutonium 238, and still 6600 years for plutonium 240.

To achieve an absolutely "clean" reactor, one would have to use one of the other possible fusion reactions listed in Table 1.1, thus avoiding having tritium and lithium in the fuel cycle. However, one will have to first demonstrate the proof of principle of a working fusion reactor using the deuterium-tritium cycle before considering reactors based on other nuclear reactions. For more examples, see Duderstadt and Moses (1982); Martinez-Val et al. (1993). Note that the total energy released in these fusion reactions determines the energy output. However, for the self-ignition of the capsule, only the energy contained in the charged particles is available.

Now that we have seen what fusion reaction to take, we can address the next problem: here on Earth, one has to achieve confinement in a much smaller space and much shorter time than in stars. As mentioned above, using a fusion reaction for an energy-producing system requires a huge number of such fusion reactions per second. This means one has to keep the nuclei relatively close together and prevent the plasma from flying apart

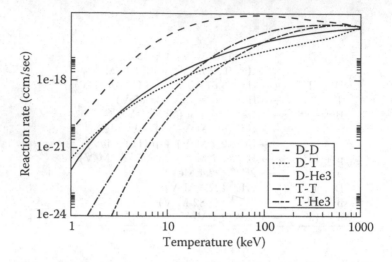

Figure 1.3. Reaction rate as a function of temperature for various fusion reactions assuming a Maxwellian velocity distribution.

by confining it long enough that a sufficient number of fusion reactions can take place.

Assuming that the plasma consists of deuterons and tritons of density $n/2$ each, the rate of fusion processes W in such a hot dense plasma state is given by

$$W = \frac{n^2}{4} \langle v\sigma \rangle, \qquad (1.8)$$

where v is the relative velocity of the two nuclei and σ the fusion cross-section. The particles in the plasma have Maxwell-Boltzmann distributed velocities with an average kinetic energy of $E_k = 3k_B T/2$. The fusion cross-section σ depends strongly on the relative velocity of the fusing nuclei and is obtained by averaging $v\sigma$ over all possible relative velocities. Figure 1.3 shows the reaction rate for various fusion reactions as a function of the temperature. Note that the temperature is expressed in energy units by multiplying its value in degrees Kelvin by the Boltzmann constant, as is common practice in this field of physics. Figure 1.3 demonstrates that at all temperatures the deuterium-tritium (DT) reaction gives the largest contribution to energy yield and is therefore the easiest reaction route.

How much energy can be produced in such a confined plasma? The energy produced per time τ depends on the kinetic energy Q of the reaction where Q is given in MeV. The ultimate aim in ICF research is an energy-producing reactor. Therefore the energy obtained from the fusion processes has to be greater than the energy to heat the plasma to such high temperatures. Or in other words, energy will be gained from an ignited DT-plasma only if this energy is larger than the total kinetic energy of all the particles. Because the kinetic energy E_{kin} of the nuclei and electrons is $E_{kin} = 3nk_BT$, it follows that only if

$$3nk_BT < \frac{n^2}{4}\langle v\sigma\rangle\tau Q,$$

the fusion reactions actually release more energy than is required to produce the plasma of such temperature and density. Re-expressed as

$$n\tau > \frac{12k_BT}{\langle v\sigma\rangle Q}, \tag{1.10}$$

this relation is called *Lawson criterion* (Lawson, 1957), which is one of the fundamental relations of confinement fusion.

In addition to the problem of confinement, the fusion particles have to have enough kinetic energy for a sufficient number of fusion reactions to take place. For DT fuel this implies a temperature of approximately 5 keV. In the case of a DT reaction with $Q = 17.6$ MeV and an operating temperature of the reactor of about 5–10 keV, the Lawson criterion becomes

$$n\tau \simeq 10^{14} - 10^{15} \text{ s cm}^{-3}, \tag{1.11}$$

where n is the number of particles per cm^3 and τ the confinement time.

1.3 The Two Approaches — Magnetic vs. Inertial Confinement

As stated before, for enough fusion reactions to take place: the plasma must be kept together at a high temperature for a sufficiently long time. Essentially two methods have been pursued in the quest for a viable fusion reactor — *magnetic confinement* (MCF) and *inertial confinement* (ICF), which aim to fulfill the Lawson criterion in two different ways. MCF tries to confine the plasma at low densities for the relatively long times of several seconds — whereas ICF yields to achieve extremely high densities for a very short time. Table 1.2 gives a comparison of the confinement times and densities in the two approaches.

Confinement time τ/s	10	10^{-11}
Lawson criterion $n_e\tau/\mathrm{s\ cm^{-3}}$	10^{15}	10^{15}

This book is devoted to the subject of ICF and only a very short description of magnetic fusion is given here. The interested reader is referred to books specializing on the subject of MCF, as for example, Braams and Stott (2002); Hazeltine and Meiss (2003).

Magnetic Fusion

Because of the required high temperature of the plasma, it cannot simply be confined in a material vessel since any contact with the walls would lead to rapid cooling. As its name implies, MCF is based on the fact that plasma can in principle be confined by applying a suitable magnetic field. This is only possible because the particles in the high temperature plasma are all charged. Magnetic fields force the charged particles of the plasma into helical orbits which follow the field lines (see Fig. 1.4). Particle movement perpendicular to the field lines is restricted while the particle moves freely in the longitudinal direction. In this way contact with the walls can be largely avoided. Because charged particles follow curved trajectories, the idea is that a suitable configuration of magnetic fields can be found so that the particles stay on closed orbits and never escape.

The objective of a closed orbit is most easily fulfilled by ring-shaped magnetic fields. However, in such a configuration, the field strength decreases with radius, which leads to a radial velocity component and a drift of the particles towards the outside. To confine the plasma for a long time,

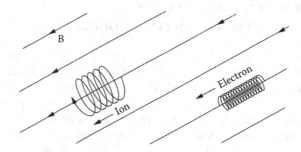

Figure 1.4. Helical movement of electrons and ions along magnetic field lines.

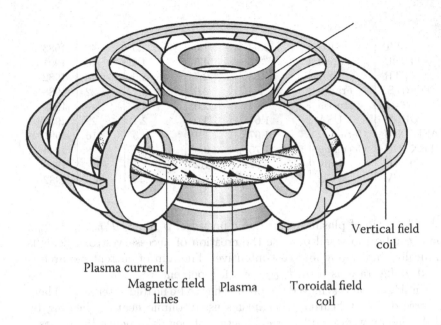

Vertical field
coil

Plasma current

Magnetic field   Plasma   Toroidal field
lines   coil

Figure 1.5. Schematic picture of magnetic configuration in a tokamak (Courtesy of IPP, Garching).

the field lines have to be twisted in such a way as to yield an absence of any radial field component. There have been numerous suggestions for such magnetic devices. The most promising ones are based on the concept of a torus-shaped steel container. The field lines create field surfaces, which lie inside one another similar to the rings of a tree. On their path through the plasma the charged particles stay on such a field surface without feeling any net radial field component (see Fig. 1.5).

The magnetic device contains a vacuum into which a mixture of deuterium and tritium is injected. The magnetic field is produced by passing an electric current through coils wound around the torus. The plasma current creates a poloidal magnetic field and the two fields combine to provide a magnetic field as shown in Fig. 1.5. This device is called a tokamak — the most common magnetic fusion configuration to date. The major magnetic fusion facilities of such a tokamak design are listed in Table 1.3.

The temperature of 10^8 K, necessary to fulfill fusion conditions, creates a plasma pressure (5–10 bar) which has to be balanced by the magnetic

JET	EU	2.96	7.0	3.5	42	1983
JT-60U	Japan	3.2	4.5	4.4	40	1991
TFTR	USA	2.5	2.7	5.6	40	1982
TORE S.	France	2.4	2.0	4.2	22	1988
T-15	Russia	1.4	2.0	4.0	–	1989
DIII-D	USA	1.67	3.0	2.1	22	1986
ASDEX-U	Germany	1.67	1.4	3.5	16	1991
TEXTOR	Germany	1.75	0.8	2.6	8	1994
FT-U	Italy	0.92	1.2	7.5	–	1988
TCV	CH	0.67	1.2	1.43	4.5	1992

field. The ratio of plasma to magnetic pressure, $\beta_{MCF} = P_{plasma}/P_{magn}$, should not be too small because the creation of necessary strong fields is technically challenging and cost-intensive. The aim of current research is to find configurations with β_{MCF} of a few percent.

Until ignition is achieved the plasma has to be heated externally. There are three different heating mechanisms used: ohmic heating, heating by high-frequency waves, and by the injection of beams of neutral particles.

Ohmic heating schemes work the following way: the particles in the plasma collide setting up a plasma resistivity. The desired heating is produced via this resistivity when current is passed through the plasma. However, the resistivity of the plasma decreases with increasing temperature and can therefore only be used in the initial heating phase of the plasma. Afterwards the other heating mechanisms must be employed.

Heating by high-frequency waves exploits the fact that in the magnetic field there exist various eigenmodes for the ions and electrons of the plasma. Radiation with electromagnetic waves of a matching frequency can lead to resonances. The particles extract energy from the wave field, resulting in a high collision rate of the hydrogen nuclei. The specific eigenmode that is used for heating exploits the circular motion of the electrons and ions around the field lines. It can be used for heating by high-frequency waves. The typical frequency (cyclotron frequency) at the relevant magnetic field strength is 10–100 MHz for the ions and 60–150 GHz for the electrons.

A third method to heat the plasma is the injection of neutral particles with an energy of several 10 keV. As the particles enter the plasma, they become ionized by collisions. Caught as fast particles in the magnetic field, they release their energy by interacting with the plasma in a relatively short time.

When the plasma is heated to sufficiently high temperatures, it will

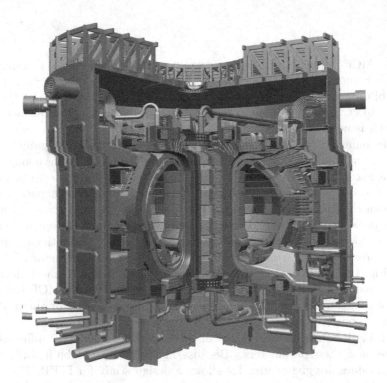

Figure 1.6. Cutaway of ITER tokamak design study (published with permission of ITER).

ignite and α-particles and neutrons are produced. Whereas the α-particles are stopped in the plasma, the neutrons easily penetrate the walls because they are unaffected by the magnetic fields. In this way the α-particles provide additional plasma heating while the fusion process keeps running, whereas the neutrons enter the blanket of absorbing material surrounding the torus. If the confinement were ideal, this process could go on until all the fuel is used up.

However, there are two processes that work against indefinite confinement. The first is collisions in the plasma: although they are necessary for the fusion process to take place, in the long run they destroy the confinement. When two particles collide the particles are temporarily disconnected from their magnetic field line and move to neighboring ones. After many collisions the particles can move from a central position to the outside and eventually hit the torus wall. The second process that can destroy the confinement is plasma instability. An instability occurs if an initially small perturbation induces a further disturbance, which in turn increases the

but in MCF devices a rather large variety of these instabilities can occur, always reducing the quality of the confinement. For more information on instabilities in MCF, we refer the reader to the specialized literature on this subject, and we will see in the following chapters that instabilities are a major issue in ICF research, too.

In summary, the problem is to keep the effects that work against confinement as small as possible to try to sustain the plasma long enough to give a net energy gain. Considerable progress has been made to achieve this objective since the early days of magnetic confinement experiments. Whereas the confinement times in 1955 were of the order of 10^{-5} s, plasmas can now be kept together for several seconds. The Joint European Torus (JET) experiment demonstrated that magnetic fusion is scientifically possible. Now the International Thermonuclear Experimental Reactor (ITER) (see Fig. 1.6) is planned as the next step magnetic fusion research device. The immediate aim is to satisfy the Lawson criterion with a MCF reactor with a confinement time of about 10 s and a plasma density of about 10^{14} particles per cm^3.

Table 1.3 shows a current list of the 10 biggest magnetic confinement devices and their parameters. Because ITER is still not built, only the design values are given: Fig. 1.4 shows a design study for ITER. Because the cost for building such research facilities increases dramatically with size, it is only possible to build such devices through a collaboration of several countries. At the moment ITER is planned to be built as an international consortium. ITER will not yet be a plant for generating electricity, but its main purpose is to study the engineering feasibility of a future MCF reactor. The next step would than be an actual reactor, the Demonstration Fusion Reactor (DEMO), which is already in its planning stage.

The Basic Ideas of Inertial Confinement Fusion

In contrast to MCF, which tries to confine the plasma at low densities ($\sim 10^{14}$ to 10^{15} cm^{-3}) for several seconds, ICF goes a different route to fulfill the Lawson criterion. Here, the confinement times are extremely short ($\leq 10^{-10}$ s), but the particle densities are typically greater than 10^{25} cm^{-3} (see Table 1.2). In this scheme a small amount of fusionable material is compressed to very high densities and temperatures by applying strong external forces.

This is done by using a capsule consisting of a spherical shell filled with deuterium-tritium gas (≤ 1.0 mg/cm^3). The shell itself consists of a high-Z material at the outside and an inner region of DT, which forms the bulk

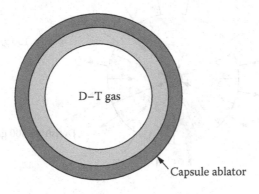

D–T gas

Capsule ablator

Figure 1.7. Schematic picture of target capsule.

of the fuel (see Fig.1.7). To reach the conditions of high temperature and density required for fusion, the capsule needs to be exposed to an enormous burst of energy applied as symmetrically as possible. The required energy input to drive this whole process is very high: to heat a 1-mm diameter capsule of fuel to 10 keV temperatures requires 10^5 J, which can be supplied by intense laser light or ion beams. This might not seem so demanding, but the energy has to be to delivered in a few picoseconds to the outer part of the target shell. Because of this burst of energy onto the outer part of the target shell, the shell heats up, ionizes and vaporizes immediately — this process is called *ablation*.

As the outer part of the shell blasts off, the inner part — essentially the fuel — is strongly accelerated toward the center of the sphere as a consequence of momentum conservation. In some sense the capsule behaves like a spherical ablation-driven rocket (this analogy will be described in detail in Section 5.3). As the fuel implodes toward the center of the capsule, it is compressed to high densities and thermonuclear temperatures. The compression shock wave drives the fuel to reach a density of several hundred gram cm^{-3} and fusion-ignition temperatures at the center, so that ignition can occur. As ignition is reached, the fusion energy produces an outward directed pressure that soon overcomes that of the imploding wave and the capsule blows back out in a very short time. In this way the required density and temperatures required could be achieved, but what about the confinement time?

The confinement time of the plasma is mainly determined by the radius R of the capsule. As the inward motion is driven by a shockwave, which moves approximately at the sound speed c_s, the confinement time can be

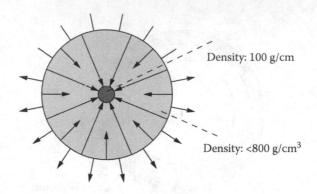

Density: 100 g/cm

Density: <800 g/cm^3

Figure 1.8. Schematic picture of hot spot concept.

roughly estimated by the ratio of capsule radius R (typically 100 μm for current target designs) over the sound speed, as $t_c \simeq R/c_s = 10^{-9}$ s. More detailed numerical simulations indicate that times of about 10–20 ns are more realistic.

In the context of ICF, the Lawson criterion is often re-expressed as a relation between the fuel density ρ and the sphere radius R. For a freely expanding sphere — where the expansion occurs with the speed of sound — the *disassembly time* can be roughly estimated by (Martinez-Val *et al.*, 1993)

$$\tau \simeq R/4c_s. \tag{1.12}$$

The number density n is related to the fuel density by $n = \rho/m$. From the Lawson criterion (Eq. 1.10), it follows that

$$n\tau \simeq \frac{nR}{4c_s} = \frac{\rho R}{4c_s m}.$$

Efficient burn requires $n\tau$ to be well above the Lawson criterion. Using $n\tau \simeq 2 \times 10^{15}$ s/cm^3 leads to a first rough estimate of

$$\rho R \simeq 3 \text{ g/cm}^2. \tag{1.13}$$

If we also take into account the fuel depletion (for the details see Section 7.3), the burn fraction Φ_b at 20–40 keV burn temperature is approximately given by:

$$\Phi_b \simeq \frac{\rho R}{6 + \rho R} \qquad [\rho R] = \text{ g/cm}^2. \tag{1.14}$$

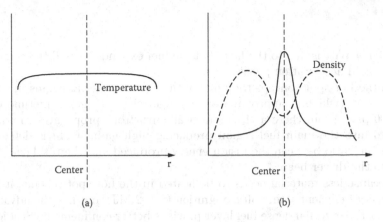

Figure 1.9. Comparison of the temperature and density profile in the central capsule area before ignition in a) the volume ignition concept and b) the hot spot concept.

The efficiency of the burn — the fusion yield E_f — for an ICF capsule is directly related to the burn fraction by $E_f = \epsilon_f \Phi_b M$, where ϵ_f is the specific energy of the fusion reaction and M the imploded fuel mass.

In the early 1960s and 1970s, people thought that inertial confinement fusion would be achieved relatively quickly, because the energy requirements for igniting the fuel did not seem too demanding (Nuckolls *et al.*, 1972). Unfortunately it turns out that not all the energy contained in the driver can actually be used for ignition. A lot of energy is lost on the way through various conversion processes from the laser to the final burn of the fusion material. This in turn means that to have the required energy for fusion, there has to be much more energy in the driver than first anticipated, and efficiency losses have to be minimized.

One important consideration is the way the fuel is compressed to high densities and temperatures. In the early days of fusion research, it was thought that the whole of the fuel should be compressed to fusion conditions at the end of the compression phase. This concept is called *volume ignition*. It turned out that this would require an unrealistically high driver energy of ~60 MJ (Cichitelli *et al.*, 1988).

The two key points here are that: (1) it takes more energy to heat fuel than to compress it and (2) the compression of hot material is more energy-consuming than for cold material. For these reasons, the so-called *hot-spot concept*, illustrated in Fig. 1.8, is considered more likely to achieve the fusion goal. In this approach the fuel moves inward with increasing velocities as the driver deposits its energy. The result of this acceleration

of the inner part is due to the fact that the fuel expands toward the center in the initial acceleration phase.

In the hot-spot concept the burn of the fusion material begins in the central area (which is approximately 1 μm in size and has a lifetime of 100–200 ps). From there a thermonuclear burn front propagates rapidly outward into the main fuel region producing high gain. Gain is defined here as the ratio between the fusion energy produced and the total energy put into the driver beams.

Because less material needs to be heated in the hot-spot scheme, it is more energy efficient than volume ignition ($\sim$1–2 MJ) and has the advantage that the external dense fuel layer provides better confinement. Studies show that if the target is constructed in such a way that the central hot-spot contains 2% of the total fuel mass, heating the hot-spot mass and compressing the remaining fuel will need comparable energy. One important issue, which will be discussed in detail in Section 4.6, is that any premature heating of the material has to be avoided as far as possible, because this would completely jeopardize the compression.

Apart from volume and hot-spot ignition, there exist additional ignition scenarios. These will be discussed in more detail in Chapter 11. Because the hot-spot ignition is the prevailing scheme for the ICF facilities under construction now, we will concentrate on this concept in the following chapters. More information about volume ignition can be found in the literature, for example in Brueckner and Jorna (1974); Kidder (1974); Bodner (1974); Meyer-ter-Vehn (1982); Lindl *et al.* (1992) and Andre *et al.* (1994).

1.4   Stages in Inertial Confinement Fusion

The above picture gives only a rough outline of the ignition and burn physics of a ICF target. In reality there exist several different phases in the ICF process that will be described in a short fashion for the hot-spot concept. These different stages will be discussed in more detail in Chapters 4 and 5.

Interaction Phase

The interaction phase is the initial phase in which the energy is delivered onto the capsule containing the DT fuel. There are mainly two options for the energy input — either by beams of laser light or particle beams. For

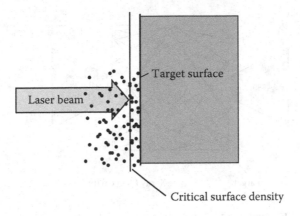

Figure 1.10. Schematic picture of the critical density formed when a laser interacts with the target.

phenomena in which only the energy input itself is of relevance, often no distinction is made between laser and particle beams: the term *drivers* is used for the energy source in the ICF context.

However, the initial interaction process differs significantly when laser or particle beams are used as drivers. Basically, laser light interacts only with the surface of the matter it encounters, whereas beams penetrate a certain distance into the material. Therefore the detailed processes of the interaction phase depend on which type of driver is used. In either case, the aim is to transfer as much energy as possible into compression energy: the pros and cons of the different driver types will be discussed in Chapters 2 and 10.

Because research with laser drivers is much more advanced in terms of achieving fusion conditions, for the moment we will assume the driver to be a laser beam. In this case, a plasma is created immediately as soon as the laser beam comes into contact with the outer surface of the capsule and expands outward from this surface. As Fig. 1.10 shows, the density of this plasma will be highest close to the capsule surface and lower further away. As soon as the plasma is created, the laser beam has to penetrate through it to reach the capsule. Now, the problem is that above a certain *critical density*, the plasma will hinder the laser beam from penetrating any further. Because the critical density surface is located at some distance from the solid target surface, the laser energy is not deposited directly onto the capsule surface any more.

The location of the critical density surface depends strongly on the wavelength, intensity, and pulse length of the laser beam. The choice of

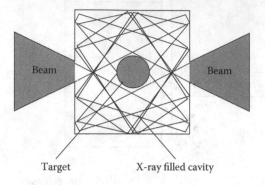

Beam

Beam

Target

X-ray filled cavity

Figure 1.11. Schematic picture of indirect drive.

these parameters is thus essential for an efficient coupling of the laser energy to the target. These parameters determine not only the gap between the critical surface and the target surface, but also the amount of ablation and the efficiency of the subsequent compression phase. The details of these dependencies will be discussed in Chapter 3.

Compression Phase

To a large extent, the interaction phase already determines how success-ful the compression phase will be (i.e., to what degree the capsule can be illuminated uniformly on its entire surface). Illumination nonuniformity occurs on two scales — microscopic and macroscopic. Macroscopic nonuni-formities can, for example, be caused by an insufficient number of beams or the existence of a power imbalance between the individual beams. One rea-son for microscopic nonuniformities is the presence of spatial fluctuations within a single beam itself. There are a number of causes for nonuniform illumination on both scales, which will be discussed in detail in Chapter 6. The important point is that both types of nonuniformities can lead to instabilities in the compression phase.

There are two ways to handle the macroscopic instabilities. The obvi-ous one is to reduce the macroscopic nonuniformities by taking a sufficient number of beams. This is done in the *direct-drive* ICF scheme. However, using many laser beams makes such systems very expensive and technically challenging. So many smaller scale direct-drive experiments are performed with just a few high-power beamlines in order to try and infer how a system with more beamlines would perform.

As an alternative to this direct approach, the *x-ray* or *indirect-drive* approach has been developed, mainly in the United States, but also in

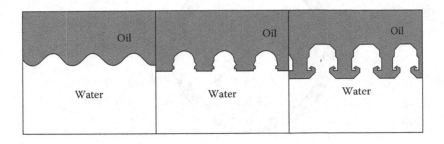

Figure 1.12. Schematic picture of the temporal development of Rayleigh–Taylor instabilities.

France, the United Kingdom and Japan. In this scheme the laser energy is first absorbed in a *hohlraum*, which is essentially an enclosure around the ICF capsule. A schematic picture of the indirect-drive scheme is shown in Fig. 1.11. Here the laser does not strike the capsule directly but instead the inside of the enclosure. This enclosure consists of high-Z material and emits x-rays when it is heated by the laser beams. It is these x-rays that then drive the implosion of the ICF capsule. Current target designs allow a conversion of the laser energy to x-rays of about 70–80%. Although this scheme therefore needs a higher energy input, it is less sensitive to hydrodynamic instabilities and the requirements on the laser beam uniformity are lower.

It is still not clear whether a direct or indirect approach would be better for producing inertial fusion energy (IFE) in a power plant and there are still vigorous experimental campaigns on both kinds of schemes.

In the past the most powerful laser systems for direct-drive experiments were the GEKKO XII at Osaka University in Japan and the Omega Upgrade at the University of Rochester Laboratory for Laser Energetics in the United States. The GEKKO XII system consisted of 12 beams that delivered 10 kJ of energy in 1 ns at either 0.5 or 0.35 μm wavelength. GEKKO XII has been redesigned for fast ignition research and plans to build a device with 60 beams have since been canceled. Today only the Omega Upgrade has 60 beams and has achieved 40 kJ of energy for different pulse shapes. However, there are plans to use the National Ignition Facility (NIF), which is under construction now and optimized for an indirect-drive scheme, to perform some direct-drive experiments as well.

The indirect-drive scheme is very much favoured by the United States and France, where military applications play a major part in the ICF pro-

Figure 1.13. The parameter R and ΔR for the definition of the in-flight aspect ratio.

gram and new laser systems (NIF and Laser Megajoule [LMJ]) are under construction. However, for a power plant, the whole process of igniting a capsule has to be repeated at a rate of seconds rather than days (as in forthcoming experiments) which might eventually make the direct-drive scheme more favourable.

Whatever the scheme, instabilities cannot be completely avoided. In particular one has to live with the occurance of the so-called Rayleigh–Taylor class of instabilities. These instabilities can occur when a denser material pushes onto a less dense one — as in the classic example of water on oil. If this metastable state is perturbed, a mixing between the two regions can set in. This is shown in Fig. 1.12. In ICF there are no heavy and light fluids, so why do Rayleigh–Taylor instabilities occur? When the targets are compressed, hot plasma pushes onto colder plasma. This is equivalent to heavier fluid pressing onto a lighter one, and Rayleigh–Taylor instabilities can also develop in this situation. A mixing of cold and hot plasma occurs, which in effect leads to an undesired cooling of the hot plasma.

This is obviously bad for the compression, so the targets have to be designed in such a way that Rayleigh–Taylor instabilities are minimized as far as possible. It turns out that the ratio of the shell radius $R(t)$ to the shell thickness $\Delta R(t)$ is the crucial parameter (see Fig. 1.13). Calculations show that this so called *in-flight aspect ratio* $R(t)/\Delta R(t)$ has to be of the order of 25–40, and not just at the beginning but at any moment during the implosion. So the need to avoid Rayleigh–Taylor instabilities directly influences the design of the deuterium-tritium containing ICF capsule.

The important parameters for the growth rate of Rayleigh–Taylor instabilities during ablation are the wavenumber of the instabilities, the ac-

description of how these parameters influence the target design, see Chapter 8.

Let us now assume that the target is designed in an ideal way to avoid Rayleigh–Taylor instabilities — what other requirements exist for the compression phase? The acceleration should be performed so that the creation of so-called hot electrons is avoided as far as possible. These hot electrons can *preheat* the fuel and create their own shock fronts. Preheating the fuel is undesirable, because, as described above, it is then more difficult to compress it.

Avoiding preheat is especially necessary if a laser is used as driver. However, to avoid preheat the pulse shape can be chosen in such a way that the appearance of undesirable additional shocks is avoided to some degree. Nevertheless, shock waves can not completely be prevented if one wants to build up the pressure in a reasonable time. Therefore a low-power prepulse is used and a succession of increasingly intense pulses can then accelerate the fuel nearly isentropically (for a detailed description see Section 2).

Deceleration Phase

When the inner part of the fuel reaches the center of the capsule the deceleration phase begins. The kinetic energy of the inner part of the fuel is converted into internal energy. The result is that both temperature and density increase in the center, whereas the main part of the fuel remains relatively undisturbed.

In the hot-spot concept, in-flight fuel velocities of at least 2×10^7 cm/s are needed to create the temperature and density necessary to ignite the plasma in the hot spot region.

To obtain this high fuel density and temperature in the hot-spot area, a succession of increasingly intense pulses is needed to achieve the necessarily isentropic compression. In the deceleration phase the last of this succession of shocks has to act at the same time as the first shock on the compressed fuel in the center. So the timing of the shocks is essential for this phase to be successful (more details in Chapter 2).

Ignition and Burn Phase

When the temperature and density conditions in the hot-spot area are right, ignition occurs. The α-particles produced deposit their energy primarily in this central area and heat it up very quickly. The radiation, the fusion

The entire process takes approximately 10 ps. During this time a very high pressure builds up that will eventually blow apart the remaining fuel and thermalized α-particles. This is then the end of the ICF cycle. In a reactor the next target has to be injected and the whole process starts all over again. Because in this last stage α-particles are produced, safety aspects are an important point, which will be discussed in Section 9.5.

Gain

In the fusion process energy is only gained if the energy deposited by the fusion products exceeds the necessary input energy. Unfortunately, this input energy is not just the energy needed to heat the fuel: several inefficiencies have to be taken into account along the way. First of all there are various losses in the driver itself because of radiation etc., this driver inefficiency results in a 3–20 factor loss of the energy initially put in the system. In addition, there are losses in the compression dynamics for example by Rayleigh-taylor instabilities and the limited burn efficiency mentioned above. The implosion inefficiency leads to a 10–20 factor higher requirement for the input energy.

These other losses of energy during the ICF process are the real problems in achieving viable fusion, so high driver efficiency and high energy gains are the crucial points. The former is a question of technological development, the latter requires sophisticated target design and tight beam specifications. How far are we from the goal of achieving fusion with ICF?

Status

Fusion conditions have not been achieved to date, but two major milestones have been reached independently: bursts of 2×10^{14} neutrons have been measured (Soures *et al.*, 1996) and 600 times liquid density ($\sim$120 g/cm^3 and $\rho R \sim 0.1$ g/cm^2) of compressed fuel has been achieved (Yamanaka, 1989a). In the experiment with the high neutron yield the temperature was 15 keV and the obtained density only 2 g/cm^3. In the high-density, experiment the maximum temperature was around 300 eV, far too low to produce an appreciable number of fusion reactions.

It might seem strange that these records in density and neutron production lie some time back. The reason is that the reachable density and neutron production are mainly limited by the energy provided by the laser. So new records can only be expected when the construction of the next generation of lasers is completed. The completion of NIF and LMJ will

1.5 Outline of the Book

The problem with writing a book about inertial fusion is that nearly everything is interconnected, so that an ideal way of structuring the material is not apparent. The same is true for learning how ICF works — all stages of ICF are strongly intertwined and a lot of cross-referencing is required. In this book I have taken a chronological approach, following the temporal development of the inertial confinement process. Therefore, unusually for a book on ICF, the material starts with the laser which provides the pulse(s). A short introduction to the basic principles of lasers will be given before we concentrate on the high-power lasers necessary for inertial confinement fusion.

From the time the laser hits the capsule, we are dealing with matter in a plasma state, so Chapter 3 gives an overview of basic plasma physics required to understand the physical processes in ICF. In Chapter 4 the interaction of the laser with the target is described. This tackles the question how the laser energy is deposited in the target. The way this energy is then used to drive the compression that leads to the burn is described in Chapter 5. In Chapter 6 we deal with the instabilities that make the processes described before less efficient, and in Chapter 7 we discuss the energy requirements and the expected gain in ICF.

In Chapter 8 the implications of the interaction physics on the target design are investigated. In Chapter 9 we discuss what would have to be considered in addition for a future energy-producing reactor. In Chapters 10 and 11 alternative routes to fusion will shown, namely heavy-ion driven fusion and the so-called fast ignition concept. Finally in Chapter 12 a glossary of the terms used in ICF is given — the ABC of ICF.

Chapter 2

Laser Drivers for Inertial Confinement Fusion

The whole inertial confinement fusion (ICF) process starts with the creation of beams in the driver. Lasers are by no means the only option as drivers for ICF. In fact it is believed that for energy-producing ICF power plants, heavy-ion beams would be much more suitable than lasers (see Chapters 9 and 10). The advantage of heavy-ion beams would be a high repetition rate for the driver delivering its energy into the target, and a much better efficiency. The driver efficiency η_{driver} is defined as the ratio of "wall-plug" electrical power E_{el} that has been converted to driver energy E_{driver},

$$\eta_{driver} = \frac{E_{el}}{E_{driver}}. \tag{2.1}$$

Heavy-ion drivers have a 2–4 times higher efficency than lasers. However, heavy-ion beam facilities are still far from delivering sufficient energy onto a target. Laser systems are much more advanced in this respect and it is likely that the first inertial fusion reactor will still use a laser driver. Therefore we will restrict the discussion in this chapter to lasers. However, in Chapters 9–11, different driver options such as heavy-ion and the so-called fast ignition scheme will be considered in detail.

In this chapter we will first give a very compressed overview of the basics of lasers physics in general and will then describe the relevant facts for lasers as ICF drivers. Because this can in no way be a complete description of this highly complex subject, the interested reader is referred for example to Davis (1996) for more details.

2.1 Basics of Laser Physics

Lasers differ from other radiation sources by being

- monochromatic,

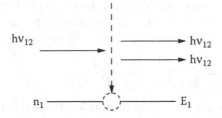

Figure 2.1. Schematic picture of spontaneous emission between energy levels E_1 and E_2.

- spatially coherent,
- temporal coherent, and
- of high brightness.

The basic reason why lasers are such powerful light sources is that the energy input usually takes relatively long (for a neodymium laser $\sim$1 ms) whereas the energy is released on much shorter timescales ($\sim$1 ns) giving a 10^6 increase in power. This happens the following way: energy is pumped into the laser medium and the atoms inside are excited to higher energy levels. As they decay they emit photons, which might hit an other excited atom. This atom then emits a photon exactly in phase with the first one — this process is called stimulated emission(see Fig. 2.1). This whole process can repeat itself again and again, leading to an amplification of light, where all photons travel in the same direction in phase and eventually can form a beam. Such a beam can then be focussed to very high irradiances.

Now we consider this mechanism for producing laser light in more detail: assuming a laser system with only one resonant frequency between the levels of energy E_1 and E_2, an excited atom can decay to a lower energy level either spontaneously or by stimulated emission (Einstein, 1917). Assuming a system of atoms in thermal equilibrium with electromagnetic radiation at a temperature T, the individual atoms can either absorb or emit photons with an energy

$$\hbar\omega = E_2 - E_1, \tag{2.2}$$

where E_1 and E_2 indicate the energy levels 1 and 2, with level 1 being lower than level 2 — see Fig. 2.1. All together the photons radiate a black

where $U_p(\omega)d\omega$ is the radiation energy within the frequency interval $[\omega, \omega + d\omega]$ per unit volume. The numbers of atoms n_1 and n_2 per unit volume at level 1 and 2 in the system are related by the Boltzmann distribution

$$\frac{n_2}{n_1} = \frac{g_2}{g_1} \exp\left[\frac{-\hbar(E_2 - E_1)}{k_B T}\right],$$

where g_1 and g_2 are the degeneracy of the atomic levels 1 and 2, respectively. The rate equations

$$\frac{dn_1}{dt} = +An_2 + B_{21}U_p(\omega)n_2 - B_{12}U_p(\omega)n_1$$

$$\frac{dn_2}{dt} = -An_2 - B_{21}U_p(\omega)n_2 + B_{12}U_p(\omega)n_1$$

describe the change of the populations on levels 1 and 2, where A, B_{12}, and B_{21} are known as Einstein coefficients. They are defined by the relations

$$B = B_{21} = B_{12}\left(\frac{g_1}{g_2}\right)$$

$$\frac{A}{B} = \frac{\hbar\omega^3}{\pi^2 c^3} = U_p(\omega) \exp\left(\frac{\hbar\omega}{k_B T}\right) - 1. \qquad (2.4)$$

In thermal equilibrium the populations are constant, therefore dn_1/dt and dn_2/dt are zero. For $\hbar\omega \ll k_B T$ the number of stimulated emissions is much smaller than the number of spontaneous emissions; therefore, it follows from Eq. 2.4 that

$$U_p(\omega) = \frac{A}{B} + 1,$$

whereas in the case $\hbar\omega \gg k_B T$, Eq. 2.4 reduces to

$$U_p(\omega) \exp\left(\frac{\hbar\omega}{k_B T}\right) = \frac{A}{B}.$$

In contrast to lasers, normal light bulbs have a relatively small temperature so that spontaneous emission dominates, which in turn means that the optical spectrum is incoherent.

One can define the transition probability per unit time of the excited level as spontaneous emission lifetime

$$\tau_{sp} = \frac{1}{A}$$

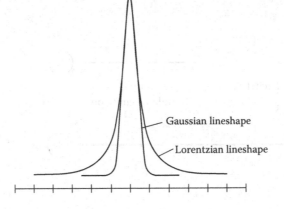

Figure 2.2. Comparison of Lorentzian and Gaussian line shapes.

and equivalently the induced emission lifetime

$$\tau_{in} = \frac{1}{BU_p}.$$

It holds for the transition probability $1/\tau$ per time unit for the excited state under consideration that

$$1/\tau = 1/\tau_{sp} + 1/\tau_{in}. \tag{2.5}$$

In general, the transition will not only be from an exited level 2 to level 1, but induced by radiation over a range of frequencies around the resonant frequency. Therefore the laser beam has not a single δ-like frequency but emits radiation according to a spectral function $g(\omega)$, which normalizes as $\int g(\omega)d\omega = 1$ for $\omega_0\tau_{sp} \gg 1$, where ω_0 is the resonance frequency. Taking this into account, the number of stimulated emissions $B_{21}U_p(\omega)n_2$ changes to

$$n_2 \int B_{21}U_p(\omega)g(\omega)d\omega = \frac{\pi^2 c^3 n_2}{\hbar\tau_{sp}} \int \frac{U_p(\omega)g(\omega)d\omega}{\omega^3}$$

with the natural line shape

$$g_n(\omega) = \frac{2\tau_{sp}}{\pi}\left[\frac{1}{1 + 4\tau_{sp}^2(\omega - \omega_0)^2}\right]. \tag{2.6}$$

The spectral broadening of the line is described by the full width of the half maximum (FWHM) of $g_n(\omega)$ is $\Delta\omega_n - 1/\tau_{sp}$.

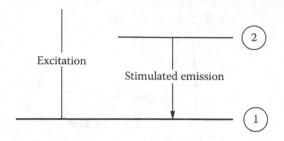

Figure 2.3. Simple three-level laser scheme.

The process described above is not the only reason for line broadening — collisions with atoms, ions or electrons in the laser medium can also contribute. If the average collision time is τ_c, the line shape broadening due to collisions is

$$g_c(\omega) = \frac{\tau_c}{\pi} \left[\frac{1}{1 + \tau_c^2(\omega - \omega_0)^2} \right],\qquad(2.7)$$

with a FWHM of $\Delta\omega_c = 2/\tau_c$. These two — natural and collisional — line shape broadening effects have Lorentzian profiles (see Fig. 2.2) and are homogenous broadening mechanisms, which means that the spectrum of each atom is broadened the same way. In contrast, inhomogeneous broadening occurs if either locally (Doppler broadening) or if inhomogeneities in the laser medium are responsible. In these cases the broadening occurs at random resulting in a Gaussian line shape (see Fig. 2.2).

The Doppler effect changes the line shape to a Gaussian, because resonance absorption at ω_0 is even possible for nonresonant frequencies ω, if $\omega = \omega_0/(1 \pm v/c)$. This leads to a line shape of the form

$$g_D(\omega) = \frac{1}{\omega_0} \left(\frac{mc^2}{2\pi k_B T} \right)^{1/2} \exp\left[\frac{mc^2}{2k_B T} \left[\frac{(\omega - \omega_0)^2}{\omega_0^2} \right] \right],\qquad(2.8)$$

with a FWHM of $\Delta\omega_D = 4\omega_0 k_B T \ln 2/(Mc^2)$. In a real laser all line broadening mechanisms may be present at the same time and the line shape will be the convolution of all different line shapes.

For lasers to work effectively, it is necessary that the excitation level 2 is more populated than level 1. This situation, where $n_2 > n_1$, is called population inversion. In thermal equilibrium the Boltzmann equation does

it is possible to achieve a population inversion between level 3 and 2 by pumping energy into the system. This energy input into the laser medium can, for example, be achieved by irradiating the laser medium with flash lamps. We will describe this process in more detail later in this section. The result of this energy input is that atoms are excited into the upper level 3. However, because the light from the flash lamp is not monochromatic, only a small proportion of the incident photons will be suitable for exciting the atoms. Therefore it is desirable that this upper level has a large linewidth with a wide frequency range so that a high efficiency in converting the light from the flash lamps into excitation of the atoms can be achieved. The excited atoms decay rapidly from level 3 to level 2. In contrast to level 3, this level 2 should have a narrow line width combined with a relatively long lifetime. This way population inversion between level 1 and 2 — the lasing transition — can be achieved.

Besides the three level scheme described above, there exists a variety of level schemes. The simplest model capable of describing the laser types of interest in ICF applications is the four-level laser illustrated in Fig. 2.4. The advantage of the four-level scheme is that this population inversion can be obtained more easily in materials that use a transition to an energy level above the ground state. The reason is that stimulated emission can start as soon as the population of the upper lasing level 3 is larger than that of the lower level 2, which is much smaller than that of the ground state 1.

The general rate equations in a four-level laser are given by

$$\frac{dn_4}{dt} = W_{14}n_1 - (W_{41} + A_{41} + S_{43})n_4$$

$$\frac{dn_3}{dt} = W_{23}n_2 - (W_{32} + A_{32})n_3 + S_{43}n_4$$

$$\frac{dn_2}{dt} = W_{12}n_1 - (A_{21} + S_{21})n_2$$

$$n_0 = n_1 + n_2 + n_3 + n_4, \tag{2.9}$$

where $W = U_p(\nu)B$ is the stimulated emission rate and S the radiationless emission rate. Just before the onset of the laser action, the states 2 and 4 are very sparsely populated and to a first approximation can be neglected. The equation system (2.9) then simplifies to

$$\frac{dn_3}{dt} = W_{14}n_1 - A_{32}n_3.$$

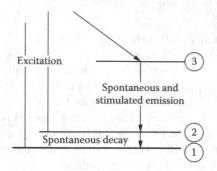

Figure 2.4. Simple four-level laser scheme.

The steady state solution of this rate equation is

$$\frac{n_3}{n_1} = \frac{W_{14}}{A_{32}}.$$

Substituting W_{14} and the lifetime against spontaneous emission $A_{32} = T_3$, it follows

$$n_3 = \frac{n_1 W_{14}}{A_{32}} = \frac{8\pi \nu^2 \eta^3 T_3}{c^3 g T_p}$$

where g is the maximum of the line width shape function and T_p the photon lifetime. The pump power required $P = W_{14} n_1 h \nu_p V$ is then

$$P = \frac{8\pi \nu^2 \eta^3 V h \nu_p}{c^3 g T_p}.$$

As the beam penetrates through a medium, its intensity changes over the traveled distance. This propagation of a monochromatic beam in a medium in direction x can be approximated by

$$\frac{dI(\omega)}{dx} = (G - \kappa) I(\omega) \tag{2.10}$$

where I is the energy flux, G its gain for a system with population inversion, and κ is a dissipation term, which describes mainly loss from collisions.

The gain is given by

$$G = \frac{\pi^2 c^2 n_r}{\omega^2 \tau_{sp}} (n_2 - n_1) g(\omega)$$

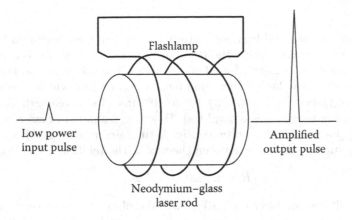

Figure 2.5. Principle of laser pumping.

and therefore depends strongly on the population inversion. Assuming that the population difference $n_1 - n_2$ is independent of the energy flux I, and G and κ are independent of x, the solution of Eq. 2.10 is

$$I(\omega, x) = I_0 \exp\left[(G - \kappa)x\right].$$

If the system is in thermal equilibrium and $n_1 > n_2$, the radiation energy decreases exponentially as it propagates. In contrast, if $n_2 > n_1$ and $G > \kappa$, the energy flux increases exponentially. This is called light amplification.

There are two classes major of high-power lasers: gas and solid state lasers, mainly distinguishing themselves through the laser medium used. In a solid state laser (Nakai, 1994), the medium is an insulated crystal or glass, and impurity ions are the active media. For high-power lasers the following laser media are currently under consideration: neodymium glass ($\lambda = 1.06\ \mu m$), KrF ($\lambda = 0.249\ \mu m$), CO_2 ($\lambda = 10.6\ \mu m$), I_2(iodine) ($\lambda = 1.3\ \mu m$), and titanium sapphire ($\lambda = 0.8\ \mu m$) with their respective wavelengths in parentheses. Neodymium lasers are the most used laser type in ICF experiments; the medium can oscillate at several lines but the line at $\lambda = 1.06\ \mu$ is mostly used as laser line.

Let us now consider as an example the individual components of a glass laser system

The mirrors are essential because a given wave has to pass back and forth through the cavity containing the laser medium many times for the oscillations to grow. A variety of oscillation modes is possible as the energy levels involved in the lasing transition have a certain line width. Because the length of the cavity is typically 10^5 to 10^6 times its wavelength, only a limited range of frequencies is amplified. However, a number of laser modes will usually be amplified simultaneously. If the mirrors have a reflectivity R_1 and R_2 and the distance between them is L, the relation

$$R_1 R_2 \exp[2L(G - \kappa)] > 1$$

has to be fulfilled for laser oscillations to take place. Stable electric field modes occur only at frequency intervals $\Delta\nu = n(c/2L)$, with $n = 1, 2, 3....$

The cavity quality factor Q (or Q-factor) describes the ability of a laser to amplify a given mode, it is defined as

$$Q = \frac{2\pi\nu_0 E_{mode}}{P_d},$$

where E_{mode} is the energy present in the amplified modes and P_d is the dissipation rate of the energy of this mode in the cavity. Those modes with the highest Q will be preferentially excited.

An oscillator contains at least two additional elements — an aperture and either a time-variable loss element or a "Q-switch." The aperture ensures that the lowest order transverse mode, which has a spatial Gaussian profile, is obtained.

The oscillator Q-switch basically functions as an electro-optical shutter changing its refraction index at certain voltages (e.g., a Kerr or Pockels cell). If this shutter is closed the oscillation growth is inhibited. During that closure time the cavity accumulates energy, because the pumping still goes on and the population inversion still increases, because very little stimulated emission depletion occurs. When this shutter is opened for a time much shorter than the build-up of the laser pulse, the switch alters the Q-factor and the ratio of the stored to the dissipated energy in the cavity changes from a high to a low value. The oscillations build up very rapidly, thus releasing the accumulated energy in a short time. The laser pulse dies out when the excited state population is depleted. By this method the Q-switch allows the laser output to be restricted to a short-duration pulse.

If the oscillator does not contain a Q-switch but instead a time-varying loss element, this loss element is modulated at a rate of $2L/c$, which is equivalent to the round-trip time of the cavity. In this way the build-up

small (10   J) to have a better control of the laser pulse. In the next step a telescope system magnifies the radius of the laser beam that leaves the oscillator before it enters the system of amplifiers.

Amplifier

The beam radiance coming from the oscillator pulse is increased by a series of amplifiers. The radiance is enlarged by a factor of 10^4 to 10^8 with pulse energies then in the range of 10 to 10^5 J.

How is this massive increase in pulse energy achieved? In the amplifier the medium has to be pumped and a population inversion achieved *before* the beam actually enters the medium. This pumping is done from the outside. In the case of a neodymium laser the population inversion can be done by using xenon flash lamps. However, the efficiency of the energy conversion from these flash lamps to gain in pulse energy is very low — 1–2%. Therefore new technological developments are urgently needed in this area. New techniques using diode lasers as pumping systems provide hope that up to 40% efficiency could eventually be achieved.

Finally, to avoid damage of the oscillators by reflection from the target of this now very powerful pulse, isolating elements have to be included in the laser design.

2.2   Lasers for ICF Applications

In the following we discuss the question what the ideal laser for ICF should look like. The most important considerations are the *laser intensity* and its *wavelength*. In contrast to many other laser applications, ICF experiments need a laser that delivers very high-power in a relatively short time ($\sim$10 ns).

The maximum electrical E_{max} and magnetic B_{max} fields of the laser in a vacuum are related to the irradiance I_L (in Gaussian units) by

$$I_L = \frac{cE_{max}}{8\pi} = \frac{cB_{max}^2}{8\pi}$$

or in practical units

$$E_{max}\left[\frac{\text{V}}{\text{cm}}\right] \simeq 2.75 \times 10^9 \left(\frac{I_L}{10^{16}\text{W/cm}^2}\right)^{1/2}$$

intensities the absorption of the laser light is proportional to the inverse of the laser intensity (Drake, 1988; Yamanaka, 1989c), so that increasingly a smaller portion of the laser light is absorbed. An additional limiting factor is the damage to the final mirror. This has to be avoided, because it would be impractical to replace the final mirror after only a few shots. The ideal compromise is an intensity of 10^{14} to 10^{15} W/cm^2. Usually pulsed lasers are used, because photon storage in the optical systems would otherwise be extremely difficult.

The second consideration when choosing a laser for ICF is the wavelength. In early ICF research a wavelength of first 10.1 μm, then around 1 μm was favored. Nowadays, it seems that a wavelength in the region of 0.3–0.5 μm is more suitable. Extensive experiments have shown that at shorter wavelengths, the favorable collisional processes are enhanced, whereas undesirable effects such as resonance absorption and Raman scattering are reduced.

The right beam intensity and wavelength are important but not the only requirements for ICF lasers. The beam also has to have a certain shape. The beam shaping occurs in three independent domains:

- Temporal shaping
  This is especially important in the hot-spot concept. To achieve the hot spot in an energy-efficient way (this will be described later in Chapter 5), the laser pulse has to deliver energy onto the target with a special time-history.

- Spectral shaping and beam smoothing
  Elimination (or at least considerable reduction) of bright and dark spots in the beam especially at the focus is essential. This is done by changing the focal beam pattern with fast changes on the wavelength scale.

- Spatial shaping
  The amplifiers are more efficient in the center than at the edges of the beam. So an initially evenly distributed beam will be weaker at the edges after passing the main amplifiers. To counteract this effect, the initial square beam is made more intense at the edges in the preamplifiers, so compensating for the higher gain in the center of the amplifiers.

Assuming that one has fulfilled all these requirements, there still remains the problem that the target is normally irradiated by several beams at

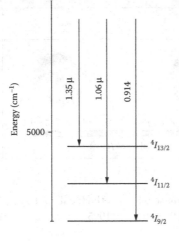

Figure 2.6. Lasing transitions in a Nd-laser.

once. This would necessarily lead to some irradiation nonuniformity; yet, irradiation uniformity is one of the main prerequisites for reaching ignition conditions. This in turn imposes strict requirements on the quality of single beam and on coordinating of the different beams too. For the targets currently under consideration for reactors, beam uniformity, power balance, and beam synchronization of $\simeq 1\%$ are required.

To summarize, the laser driver starts with a low-power laser pulse that is then smoothed, shaped, and strongly amplified. This initial weak laser pulse has only a few nJ and a beam diameter of a few μm. This pulse acquires a temporal shape and is broadened over multiple colours. Each of these pulses is amplified and shaped in preamplifier modules.

In two steps the pulse is amplified to some 10 J. During this process the beams become spatially spectrally and temporally shaped. This includes beam smoothing as well. Afterwards these beams are amplified to the intensities required to achieve fusion conditions, keeping their spatial, spectral and temporal appearance the same.

2.3 Nd-glass Lasers for ICF

At the moment glass lasers are the most advanced type of laser for ICF, based on the fact that they are able to release high-power in a short time. Indeed Nd-glass lasers have mainly been used in experiments related to ICF so far. The solid state lasing material consists of neodymium ions embedded in a matrix of yttrium aluminum garnet ($Y_3Al_5O_{12}$) crystal or glass. It is a four-level lasing system similar to the one described in Section 2.1. There are several possible transitions that could be used in

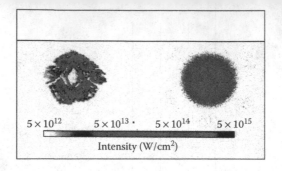

5×10^{12} 5×10^{13} · 5×10^{14} 5×10^{15}

Intensity (W/cm^2)

Figure 2.7. The effect of a phase plate on the focal spot for a Nova beam (adapted with permission from Lindl (1995), ©1995, American Institute of Physics.)

the laser medium (see Fig. 2.6). However, the infrared transition at 1.06 μm is most used with a double (0.53 μm) or triple (0.35 μm) frequency multiplication (Singh, 1987) for ICF experiments. The main laser systems under construction now will use the triple frequency of 0.35 μm, because the undesirable types of laser-plasma interactions are less pronounced at these shorter wavelengths. However, recently the idea of using 2ω light in National Ignition Facility rather than 3ω light has become popular again because one obtains about twice the energy in 2ω light. The idea is to trade the higher energy for lower coupling efficiency or symmetry.

The main aim in glass laser research has been to improve beam quality to achieve a high-laser radiation uniformity on the target. If the irradiation of the target is not perfectly uniform, it imprints on the target surface and perturbs the spherical symmetry which can seed hydrodynamic instabilities (see Chapter 6).

The number of beams and the power and energy balance among the beams determines irradiation nonuniformity at low spatial frequencies, whereas the quality of the individual beams determines the high spatial frequency irradiation nonuniformities. It has been demonstrated by Yamanaka (1989b) that the uniformity of the irradiance can be improve significantly by the following techniques:

- random phase plates (RPP)
- spectral speckle dispersion (SSD)
- polarization smoothing
- multibeam overlap (Regan *et al.*, 2005)

RPPs modify the focal-plane intensity distribution to an ensemble of fine-scale speckles with well-defined statistical properties (Obenschein,

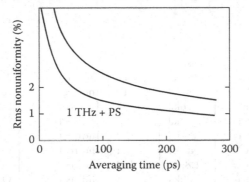

Figure 2.8. RMS non-uniformity as a function of time-calculated assuming multiple beam overlap on a spherical target in the OMEGA 60 beam geometry. Curves are shown for two different configurations of spectral speckle dispersion and polarization smoothing (from McCrory *et al.*, 2001).

1986; Lehmberg, 1987; Y. Lin, 1996). The characteristic size of the speckles depends on the diffraction limit of the focusing optics, where one important parameter is the ratio f_{sp} of the focal length to the beam diameter. The characteristic values for the speckles width d_{sp} and length l_{sp} should be approximately be given by (Rose and Dubois, 1994; Watt *et al.*, 1996; Garnier, 1999)

$$d_{sp} \sim f\lambda_0,$$
$$l_{sp} \sim 7f^2\lambda_0.$$

The technique of SSD (Rothenberg, 1997; Regan, 2000) significantly reduces irradiation nonuniformity by rapidly shifting the laser speckle pattern generated on the target by distributed phase plates. A high-frequency electro-optic phase modulator produces a time-varying wavelength modulation that is angularly dispersed by a diffraction grating.

Polarization smoothing (Tsubakimoto *et al.*, 1993) splits each beam into two components and recombines the parts so that their irregularities interfere destructively. PS instantaneously smooths the spatial beam structure, and thus can be more effective than temporal smoothing techniques.

The combination of random phase plates with the other above mentioned techniques allows a much improved irradiance uniformity of the individual beams (see Figs. 2.7 and 2.8).

Planned and Constructed Laser Systems for Ignition

An overview of the largest past, existing and planned facilities of this type are given in Table 2.1. Obviously this list is not complete, because there

Gekko XII	Japan	20 kJ	12	Direct	Modified to PW
NOVA	USA	40kJ	10	Indirect	Shut down
Phebus	France	6kJ	2	Indirect	Shut down
SG-I	China	8kJ	8	Indirect	Operating
SG-II	China	30 kJ	32	Indirect	Constructing
Beamlet	USA	40 kJ	4	Indirect	Operates
NIF	USA	1.8MJ	192	Indirect	~2011
LIL	France	60kJ	8	Indirect	Operates
LMJ	France	1.8MJ	240	Indirect	~2012
Gekko-PWM	Japan	100TW		Fast ignition	Operates
Gekko-PW	Japan	1PW		Fast ignition	Planned
Vulkan	UK	2.6kJ		Physics	Operating

are many more smaller laser systems used for ICF experiments, too. At the moment the most powerful operating laser systems for ICF research is the OMEGA system in Rochester, New York. Their main activity is experiments for direct-drive applications. In addition, 40 beams of the 60-beam OMEGA system can be rearranged to a geometry suitable for indirect-drive experiments, but the OMEGA system is not able to achieve ignition conditions in either configuration.

High-power Nd-glass lasers are under construction at Lawrence Livermore National Laboratory (LLNL) in the United States (NIF) and in France (Laser Megajoule — LMJ) with the goal of achieving ignition and burn (Lindl and McCrory, 1993). Thus it can be expected that the first ICF ignition will be obtained using a laser of the Nd-glass type.

NIF and LMJ are very similar in design but differ in the details. The original design was mainly based on the results of the indirect-drive experiments with the NOVA system, which meanwhile has been dismantled. The conceptual design of NIF can be seen in Fig. 2.9 [1] and of LMJ in Fig. 2.10.

The NIF is planned as a 192-beam Nd-glass laser system working at the tripled frequency of $\lambda = 0.35$ μm. An on-target energy of 1.8 MJ and 500 TW delivered in 20 ns is expected. Further design parameters are given in Table 2.2. The design parameters of the laser are for an indirect-drive scheme with a target capsule similar to that in Fig. 1.11. One expects that such a capsule would require a laser pulse of 1.35 MJ to be driven at a

---

[1] Here and in the following ©LLNL indicates that credit must be given to the University of California, Lawrence Livermore National Laboratory, and the Department of Energy under whose auspice the work was performed, when this information or a reproduction of it is used.

Pulse simultaneity	<30 ps
spotsize at entrance hole	500 μm

temperature of 300eV.

The main components of one such beam line, where the low power laser pulse is smoothed, shaped and amplified and eventually directed to the small target area, are shown in Fig. 2.11. At LLNL one such beamline (called Beamlet) is already constructed to test the single components of the system, similarly there exists already the small scale eight-beam prototype of LMJ called Ligne d'Intégration Laser in France. Each laser pulse must travel about 450 meters, bouncing off the equivalent of 54 mirrors and going through 2 meters of glass before it reaches the target. This whole journey

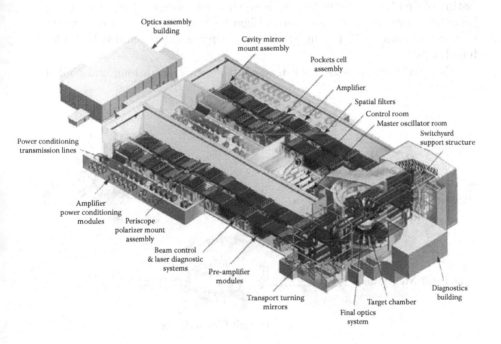

Figure 2.9. Schematic picture of the National Ignition Facility (NIF) ©LLNL[1].

Afterwards the beam is shaped in time and broadened in spectral range before it reaches the preamplifier modules (PAM) for amplification and further beam shaping.

In the PAMs the pulse will be amplified by approximately a factor of 1 million in a first step. Afterwards this mJ pulse will be again amplified by passing the beam four times through a flashlamp-pumped amplifier. In this process the pulse will be amplified to a maximum of 22 J. To cut costs, NIF is now planned to have 48 instead of 192 PAMs as originally planned, which means four beamlines will have to share one PAM. In the PAMs, spatial, spectral, and temporal beam shaping has to take place. Now the main part of the amplification of the laser beam as shown in Fig. 2.11 starts. The nominal 1-J input pulse from the optical pluse generation system is now amplified in two stages — first in the cavity amplifier and then the booster amplifier. In this process, the spatial, spectral, and temporal features from the input beam are retained. These amplifiers are far larger than in conventional lasers. The glass slabs are 46×81 cm and 3.4 cm thick, consisting of neodymium-doped phosphate glass and weigh 42 kg each. Each beamline contains 16 of such slabs. Eight such slabs are stacked together to accommodate eight beams at the same time and surrounded by 30 kJ flashlamps.

NIF will contain 7680 such flashlamps, each 180 cm long and cooled

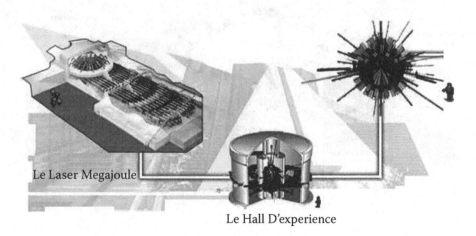

Le Laser Megajoule

Le Hall D'experience

Figure 2.10. Schematic picture of the Laser Megajoule in France ©CEA.

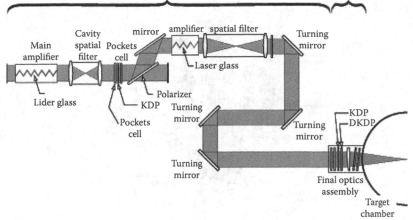

Figure 2.11. Beamline in NIF, ©LLNL[1].

by nitrogen gas. This cooling is necessary to reduce the beam distortion and be able to have a shot approximately every 8 hr. To provide the energy for the flashlamps, the energy has to be stored in a large number of capacitors. In NIF the energy will be stored in four capacitor bays, each 15 × 76 meters. These four capacitor bays will store about 330 MJ of energy. The system must provide three pulses to each of the 7680 flashlamps. The first pulse is a triggered test pulse, the second one a preionization pulse to prepare the lamps for the main charge, and, finally, the main pulse to provide the energy for the flash lamps. The pulse power is switched via huge current-capacity switches and must reliably handle 400 kA.

In NIF a Pockels cell will be used as optical switch. The electro-optic crystal is made of potassium dihydrogen phosphate (KDP). In contrast to conventional Pockels cells, which would require the crystal to be as thick as the beam diameter (here 40 cm), the so-called plasma electrode Pockels cell uses in addition a thin plate of KDP between two gas discharge plasmas. Combined with a polarizer, the Pockels cell can be used either to let the light pass through or be reflected off the polarizer. This allows the beam to pass four times through the main amplifiers before it continues its way toward the target area. However, during these four passes wavefront abberations occur within the beams caused by distortions in the amplifier glass and other optics. Just as with modern telescopes, adaptive optics is used here, the main part of which consists of a 40-cm deformable mirror. An array of actuators enables the mirror surface to be bent in such a way that it compensates for the distortions. In addition to these deformable

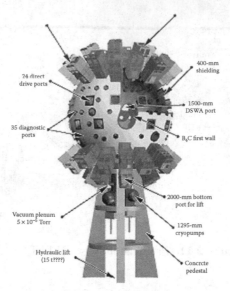

Figure 2.12. NIF focus area, ©LLNL[1].

mirrors, a computerized system is used to align and control the beams within about 30 min.

The switch-yards convert the beams that have been traveling in bundles of eight — four high and two across — to 2 × 2 arrays, which are then switched into a radial, three-dimensional configuration around the target chamber. Just before entering the target chamber, the pulses pass through the final optics assemblies (FAO), where the pulses are converted from infrared (~1.06 μm) to ultraviolet light (~0.351 μm) and focused on the target. This is done, by using two nonlinear crystal plates made of potassium dihydrogen phosphate. The first crystal converts about two-thirds of the 1.06 μm-light to the second harmonic of 0.53 μm, which is then mixed with the remaining 1.06 μm light in the second crystal to produce 0.35 μm light. The conversion efficiency is expected to be between 60–80%.

For optimal focusing each beam is focused to an elliptical spot of 500 μm by 1000 μm and has a flat-topped shape. The reason for this is to reduce laser entrance hole clearance (Lindl, 1995). The FAOs contain frequency conversion crystals, vacuum windows, focus lenses, diffraction optics, and debris shields (see Fig. 2.12). The specification for the NIF beams at this point can be found in Table 2.2.

Finally the beam will enter the target chamber (for the NIF design target chamber, see Fig. 2.13), where the laser will eventually illuminate

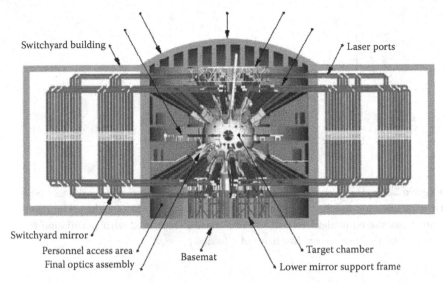

Switchyard building

Laser ports

Switchyard mirror

Personnel access area

Final optics assembly

Basemat

Target chamber

Lower mirror support frame

Figure 2.13. NIF target chamber, ©LLNL[1].

the target. This requires extremely high timing and positioning accuracy of the 192 beams. The clustering of the 192 beams in groups of 4 beams results effectively in 16 spots in each of the outer cones and 8 in the inner ones. Thus each 4 beam cluster combines to a f/8 optic.

The laser light enters the hohlraum target through two laser entrance holes in two cones. For experiments this has the advantage that the relative brightness of the two cones can be changed, so enabling asymmetries to be minimized in the energy deposition by x-ray radiation on the capsule.

One function of the target chamber is to provide a 10^{-6} torr vacuum. In addition, the target area will be temperature-controlled to 0.3°C to maintain laser positioning. The target chamber itself consists essentially of a 10-meter sphere made of aluminum alloy 5083, which is usually used to build ship structures. The sphere weighs 118,000 kg and is already installed at LLNL. After ignition it has to shield the surrounding from the emitted neutrons and gamma rays.

Present Status and Near Future

Of the large laser systems for ICF research only OMEGA-upgrade at Rochester is available for implosion experiments at the moment. It is mainly used for direct-drive experiments, but to a lesser degree indirect-drive experiments are performed as well.

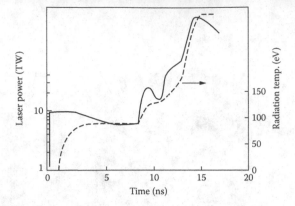

Figure 2.14. Pulse shape illustrated as the temporal development of laser power (drawn line) for a 300 eV target, which absorbs 1.35 MJ of light. The dashed line shows the equivalent radiation temperature (reprinted with permission from Lindl (1995), ©American Institute of Physics.)

The next big step in inertial fusion research will be the completion of lasers in the MJ range (i.e, NIF and LMJ). With these two laser systems it should be demonstrated that ignition in ICF can be achieved.

The schedule for NIF is the following: it is planned that the full 192 beams of NIF will be completed in 2010. The completion of the actual target chamber with its cryogenic system is estimated to take another year. The plan is to demonstrate ignition by the end of 2011. The schedule of LMJ is almost identical with the difference that no laser-plasma interaction experiments will be performed before completion of the full system.

In 2003 a beamline consisting of four beams was completed. These four beams gave 21 kJ of 1ω light, 11 kJ of 2ω light and 10.4 kJ of 3ω light each. If all beamlines would achieve the same performance, the 192-beam NIF laser would be equivalent to 4.0 MJ of 1ω light, 2.2 MJ of 2ω light and 2.0 MJ of 3ω light, exceeding the design value of 1.8 MJ and well above the 1.3 MJ the fusion considerations are based on. The beams have a shaped 25 ns pulse with a beam contrast better than 6% and a beam energy balance of better than 2%. The relative timing of the beams is 6 ps. The performance of the beamline met the criteria for beam energy, beam output, uniformity and beam-beam timing, and beam shape required for ignition.

As mentioned before, the design of LMJ is very similar to that of NIF, but it will have more beams (240). The first laser experiments are planned at 600 kJ in 2006 and with the full strength of 2 MJ in 2010. The first ignition experiments with DT fuel are envisioned for 2012.

correct. This is not only important for our general understanding of the physical processes in ICF, but essential for the planning of future facilities. The second big subject of study will everything connected with the resulting fusion products. It will be the first time that one will be able to investigate the burn physics and the material damage to the target chamber, leading the way to the next step in fusion research — the demonstration reactor.

2.4   Alternatives to Nd-glass Lasers

Despite the recent and past success of Nd-glass lasers in ICF experiments, this type of laser will probably not be used for a commercial ICF reactor. The reason is that Nd-glass lasers have a low efficiency and a low repetition rate. For example, the NIF laser has an efficency of below 1%. This means that 99% of the energy input is lost somewhere on the way; for a working reactor, such a waste of energy would be completely out of the question. The same applies for the repetition rate, as stated above for NIF a repetition rate of one shot every 8 hr is planned. At the moment it seems nearly impossible to increase the repetition rate to a few times per hour, not to speak of one shot per second as would be required for a reactor. So alternatives for a driver have to be found. Apart from completely different drivers such as heavy-ion beams, there are also possibilities of using a different types of laser drivers.

Diode Pumped Solid State Lasers

Using diodes instead of flashlamps to pump a solid state laser could allow a much higher repetition rate plus a better efficiency. For diode pumped solid state lasers (DPSSL), an efficiency of around 10% is expected. Apart from its high efficiency and repetition rate, the ultra-high brightness of the beams would make it attractive as a driver for ICF reactors.

For possible use in reactors there are three facts to consider — the achievable irradiation uniformity, the cost, and the ideal solid state material. Whether solid state lasers can be used for future ICF reactors strongly depends on the first point — the achievable irradiation uniformity on the target surface. As mentioned earlier several concepts and technologies for better radiation uniformity have been developed in recent years, SSD, RPP, polarization smoothing, and multibeam overlap. The aim is now to build high-power DPSSLs and test which irradiation uniformities can be obtained.

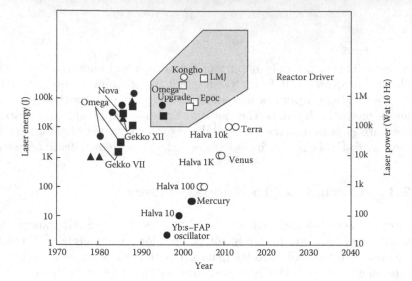

Figure 2.15. Comparison of the development of glass laser systems and DPSSL systems, where the filled symbols indicate systems that are or have been operating the the nonfilled symbols indicate the lasers under construction or planned (adapted from Nakai and Mima, 2004).

Because of steady development during the last few years, laser diodes can now meet the pumping specifications of high-power lasers, but there are still cost considerations. Current DPSSLs designs are considered far too expensive to use in a reactor. A reduction of the production cost by a factor 10–100 would be required.

Another point that requires further study is the ideal solid-state laser material. The ideal material is defined by the range of the stimulated emission cross section and the lower limit of thermal shock (Matsui *et al.*, 2000). There are different laser materials under consideration: the HALNA laser at Institute of Laser Engineering, Japan, uses Nd HAP-4, a reinforced glass material, whereas the 100 J Mercury laser at LLNL uses Yb-SFAB. For the latter, first diodes have been developed and it is under construction at LLNL. It will have a pulse length of 2–10 ns and a repetition rate of 10 Hz. The reason why LLNL chose this material is that it operates at nearly the same frequencies as Nd-glass lasers (1, 0.5, and 0.35 μm) and the hope is that the experience with Nd-glass lasers could perhaps benefit DPSSLs, especially as the target interaction should be similar.

Figure 2.15 shows the laser energy of the existing and planned DPSSLs and compares them with existing flashlamp-pumped systems. Only future research can show which of these laser materials is best suited to ICF

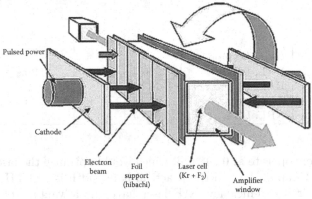

Pulsed power

Cathode

Electron beam

Foil support (hibachi)

Laser cell (Kr + F₂)

Amplifier window

Energy + (Kr + F$_2$) $\Rightarrow$ (KrF) + F $\Rightarrow$ (Kr + F$_2$) + hν (248 nm)

Figure 2.16. Main components of Electra KrF laser (Sethian et al. 2003).

reactor applications or whether other materials like optical ceramics have additional benefits.

Krypton Fluoride Lasers

There is also the possibility of not using glass lasers at all — an alternative would be excimer lasers, such as Krypton fluoride (KrF). The largest KrF laser facilities for ICF working (or planned) at the moment are summarized in Table 2.3.

As an example the main components of the Electra KrF laser at the Naval Research Laboratory are shown in Fig. 2.16. The cell is filled with a mixture of Kr, F, and argon. High-power electron beams are injected into the laser cell from opposite sides. The laser gas is isolated from the vacuum region in which the electron beams are formed by a foil support structure. The laser beam then propagates perpendicular to the electron beam. A gas recirculator ensures that the lasing medium is cool and quiescent for the next shot (for more details see, for example, Sethian *et al.*, 2003).

KrF lasers have the advantage of a much higher efficiency than Nd-glass laser — the theoretical maximum is at about 20%. This optimistic value is somewhat reduced if one takes into account the excitation energy efficiency, which then gives a more realistic efficiency of about 10%. However, these values have to be taken with some care as they have only been tested for smaller devices, tests with the Electra and Nike lasers at the Naval Research Laboratory point more in the direction of 5%. However, that would still be considerably more than that for current Nd-glass lasers.

	(project stopped)				
Sprite	RAL, UK	250 J	1	50 ns	
TITANIA	RAL, UK	1–10 kJ			
Super-Ashura	Japan	2.7 kJ	12	2.5–25 ns	
Electra	NRL, USA	500 J			5 Hz

KrF lasers operate at 0.248 μm, which would optimize the laser-target physics (see Chapter 4), and the practical bandwidth is 1–3 THz. In addition to their high efficiency, KrF laser have the advantage of a broad bandwidth, relatively easy pulse shaping, and higher pulse repetition. The latter is due to the fact that in KrF lasers the laser medium is a gas. This gas can be circulated for heat removal making a high pulse-repetition rate possible.

However, there are disadvantages as well: the pulse duration in KrF lasers is too long. In principle, this can be overcome by chopping the pulse into subpulses which are led on different paths to the target area so that they reach the capsule all at the same time. This mechanism is called multiplexing, more information on which can be found in Sullivan (1993). The shorter wavelength of KrF lasers has its disadvantage for ICF applications, too, because of the shorter wavelength the damage to the optical elements is much larger.

Only future work will show whether any of these different laser types or some other driver will be used in an ICF reactor.

Chapter 3

Basic Plasma Physics

When the laser interacts with the target surface, the atoms become immediately ionized and a plasma consisting of positively charged ions and negatively charged electrons is created. Thereafter all the processes such as absorption of the laser energy or compression of the fuel take place in a *plasma* environment. It is therefore important to consider the underlying basic physical phenomena that occur in matter in the plasma state. The reader familiar with plasma physics may skip this material and continue with Chapter 4. Only a short description of the basics of plasma physics can be given here and is also restricted to plasma phenomena essential for the understanding of the ICF process. For more information about plasma physics in general, the reader is referred to standard texts such as Chen (1984), Dendy (1994), Goldston and Rutherford (1996), and Eliezer (2002).

3.1 Debye Length and Plasma Frequency

Plasmas are not such a rare state of matter as our experience on Earth might suggest. Actually if one looks at the entire universe, 99% of all matter is in the plasma state. In nature plasmas are encountered in stars and the interstellar medium, but also closer to us in the Earth's ionosphere and in lightning. Fusion experiments aside, man-made plasmas also exist in light bulbs, discharges, flat-screen televisions, and are created in short-pulse laser experiments. Plasma conditions span a very wide range of temperatures and densities, as Fig. 3.1 demonstrates. So, for example, whereas the interstellar gas contains only a few particles per cubiccentimeter, white dwarfs have of the order of 10^{30} particles/cm^3. Similarly the temperatures range from room temperatures in nonneutral electron plasmas to 10^8 degrees K in plasmas in magnetic fusion experiments. Note that in plasma physics, as in this book, temperatures are usually expressed in the energy unit eV (for conversion, 1 eV corresponds approximately to

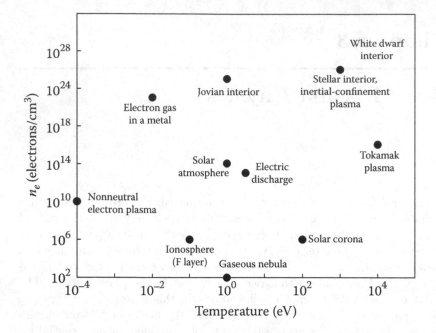

Figure 3.1. Different kinds of plasmas as a function of temperature and density.

11400⁰ K). Despite this large variety of plasmas, most of them can be treated using classical physics; only strongly coupled plasmas (upper left corner in Fig. 3.1) need special treatment, which will be described in Section 3.8.

All these different types of plasma share a common definition: a plasma is a quasi-neutral gas consisting of charged particles that show collective behaviour. Quasi-neutral means that the Coulombic forces between the particles tend to neutralize local charge imbalance in the system, which means

$$n_e \simeq Z n_i$$

where n_e is the electron density, n_i the ion density, and Z the ionic charge. Connected with this is the so-called Debye shielding effect. This describes the fact that because of the attractive and repulsive forces in the plasmas, an ion will be surrounded by electrons and vice versa (Fig. 3.2), so that on a large scale the plasma will be, on average, charge-neutral. The *Debye length* λ_D is the characteristic distance beyond which a charge is completely shielded from the effects of the surrounding charges. The Debye length λ_D

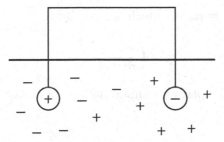

Figure 3.2. Schematic picture of Debye shielding.

is given (in SI units) by

$$\lambda_D = \left(\frac{\epsilon_0 k_B T_e}{e^2 n_e} \right)^{1/2}, \tag{3.1}$$

where ϵ_0 is the dielectric constant. It follows that for a system to be quasi-neutral, the system size has to be considerably larger than this Debye length, because Debye shielding only works if there are a sufficient number of electrons around the ion. The number N_D of electrons in a Debye sphere of radius is λ_D is given as

$$N_D = n_e \frac{4\pi}{3} \lambda_D^3.$$

The resulting potential around a shielded ion falls off exponentially and is given by

$$\Phi_D(r) = \frac{e}{4\pi\epsilon_0 r} \exp(-r/\lambda_D).$$

In thermal equilibrium the kinetic energy of the electrons has to equal that of the ions,

$$\frac{1}{2} m_e v_e^2 = \frac{1}{2} m_i v_i^2 = \frac{3}{2} k_B T_e.$$

It follows that

$$\frac{v_i}{v_e} = \left(\frac{m_e}{m_i} \right)^{1/2} \sim \frac{1}{43},$$

which means that because of their much higher mass, ions move considerably slower than the electrons. Therefore, it is the electrons that initially

plasma frequency which is given by

$$\omega_p = \left(\frac{e^2 n_e}{m_e \epsilon_0}\right)^{1/2}.\tag{3.2}$$

This plasma frequency plays an important role in almost all plasma processes.

3.2   Particle Description

There are basically two different approaches to describe a plasma — either treating it like a fluid or as many individual particles. In the latter, starting with constant electric and magnetic fields, the movement of a single particle in a plasma can be described by the Lorentzian equations of motion

$$m\mathbf{v} = q(\mathbf{E} + \mathbf{v} \times \mathbf{B}),$$
$$\mathbf{r} = \mathbf{v}t.$$

Fortunately, in the context of inertial confinement fusion (ICF) the electric field induced in the plasma is so strong that in many circumstances the magnetic field $\mathbf{B}$ can be ignored, which leaves us with a linear acceleration parallel to the electric field $\mathbf{E}$. If one would now attempt to describe the entire plasma consisting of N particles by their positions $\mathbf{r_i}$ and their velocities $\mathbf{v_i}$, this would mean solving a dynamical problem for $N > 10^{20}$ particles — obviously an impossible task. Therefore instead of the 6N-dimensional phase space $(\mathbf{r_i}, \mathbf{v_i})$ one introduces a distribution function $f(\mathbf{r}, \mathbf{v}, \mathbf{t})$ where $f(\mathbf{r}, \mathbf{v}, \mathbf{t}) \mathbf{d^3 r d^3 v}$ represents the number of particles contained in the elements (dx, dy, dz) at $\mathbf{r}$ and (dv_x, dv_y, dv_z) at $\mathbf{v}$ at the time t. If collisions can be neglected, the system can be described by the collisionless Boltzmann equation

$$\frac{\partial f}{\partial t} + \mathbf{v}\frac{\partial \mathbf{f}}{\partial \mathbf{r}} + \frac{\mathbf{F}}{\mathbf{m}}\frac{\partial \mathbf{f}}{\partial \mathbf{v}} = \mathbf{0},\tag{3.3}$$

where $\mathbf{F}$ is the force. If we substitute $\mathbf{F}$ by the Lorentz force, it follows

$$\frac{\partial f}{\partial t} + \mathbf{v}\frac{\partial f}{\partial \mathbf{r}} + \frac{q}{m}\left(\mathbf{E} + \mathbf{v} \times \mathbf{B}\right)\frac{\partial f}{\partial \mathbf{v}} = 0,\tag{3.4}$$

— the so-called Vlasov equation. This describes the motion of particles under the influence of external and internal fields. The equilibrium solution

where the thermal velocity $v_t = \sqrt{k_B T / m}$. So if collisions can be neglected, the velocity distribution is Maxwellian, just like in a gas.

However, it is not generally a valid assumption to neglect collisions in a plasma. In particular the electrons can be affected by Coulomb collisions with the much heavier ions, which for example leads to the thermalization of the system. The general Boltzmann equation includes a term $(\partial f / \partial t)_c$, which describes the effect of collisions on the distribution function

$$\frac{\partial f_\alpha}{\partial t} + \mathbf{v} \frac{\partial f_\alpha}{\partial \mathbf{r}} + \frac{\mathbf{F}}{m} \frac{\partial f_\alpha}{\partial \mathbf{v}} = \left(\frac{\partial f}{\partial t} \right)_c , \tag{3.6}$$

where $\alpha = $ e, i, n. Eq. 3.6 actually represents a system of equations one for every kind of particle contained in the system (i.e. electrons, ions, and possibly atoms), which are all connected through the collision term as $(\partial f / \partial t)_c = g(f_e, f_i, f_n)$.

The general Boltzmann equation would in principle allow the temporal change of the distribution function, including collisions between electrons and ions, to be described on a microscopic scale. But apart from very special cases, because of its complexity, only numerical solutions exist.

However, under certain conditions the collision term can be approximated. In plasma physics the approximation used most often is that of Fokker and Planck. The Fokker-Planck approximation assumes that small angle scatterings dominate: the number of collisions where the path of the particles is altered significantly is so small that it can be neglected. The Fokker-Planck term for describing the effect of collisions on the distribution function has the following form

$$\left(\frac{\partial f}{\partial t} \right)_c = -\frac{\partial}{\partial v} \left(f \langle \Delta v \rangle \right) + \frac{1}{2} \frac{\partial^2}{\partial \mathbf{v} \partial \mathbf{v}} : \left(f \left(\Delta v \Delta v \right) \right) . \tag{3.7}$$

Even after these simplifications, there is no general analytical solution, but the Fokker-Planck equation is often used to investigate kinetic plasma phenomena numerically on a macroscopic scale.

3.3 Fluid Description

If one is interested in transport processes on a global scale — for example, the overall heat transport in the initial interaction phase — it is neither necessary nor practical to investigate the plasma in such detail as to include

concern. In these cases the plasma can be considered as a single-component fluid.

To obtain the simplest one-component fluid description, one basically considers statistical averages over the velocity distribution; for example, the density is given by

$$n(\mathbf{r}, t) = \int f d^3 v.$$

In the case of a charge-neutral plasma, $n_e = Z n_i = Z n$, and the momentary state of the plasma is described by the mass density ρ rather than the distribution function f. The mass density is connected to the electron and ion density by

$$\rho = n_i m_i + n_e m_e,$$

and because the electron mass is much smaller than the ion mass, $m_e \ll m_i$, it follows that $\rho \simeq n m_i$. Similarly the velocity of the electrons and ions is combined to a single fluid velocity

$$\mathbf{v} = \frac{1}{\rho} \left(n_i m_i \mathbf{v}_i + n_e m_e \mathbf{v}_e \right) \simeq \mathbf{v}_i.$$

In hydrodynamics, the description of fluids is based on the three conservation laws of mass, momentum, and energy, resulting in a set of three differential equations for the temporal development of the fluid.

Similarly, assuming that the distribution $f(\mathbf{v})$ is isotropic, after some lengthy derivation (for which the reader is referred to plasma physics texts such as for Eliezer, 2002), the following set equations for the fluid description of a plasma is obtained:

$$\frac{\partial \rho}{\partial t} + \nabla (\rho \mathbf{v}) = 0 \tag{3.8}$$

$$\rho \left[\frac{\partial v}{\partial t} + (\mathbf{v} \cdot \nabla) \mathbf{v} \right] = \nabla \mathbf{J} \times \mathbf{B} - \nabla P + \frac{\rho}{m} \mathbf{F} \tag{3.9}$$

$$\frac{m}{ne^2} \frac{\partial \mathbf{J}}{\partial t} = \mathbf{E} + \mathbf{v} \times \mathbf{B} - \frac{1}{ne} \mathbf{J} \times \mathbf{B} + \frac{1}{ne} \nabla P_e - \eta \mathbf{J} \tag{3.10}$$

where $\mathbf{J}$ is the current density defined by the sum of the electron and ion currents $\mathbf{J} = \mathbf{J}_e + \mathbf{J}_i$,

$$\mathbf{J} = n_i Z e \mathbf{v}_i - n_e Z e \mathbf{v}_e \simeq e n (\mathbf{v}_i - \mathbf{v}_e)$$

$$\nu_{ei} = \frac{}{2^{1/2}(4\pi\epsilon_0)^2 m^2 v_t^3} \qquad (3.11)$$

with the Coulomb logarithm $\ln \Lambda = \ln(9N_D/Z)$. For a derivation of the collision frequency and the Coulomb logarithm we refer the interested reader again, for example, to Eliezer (2002).

This equation system 3.8–3.10 describes the plasma in the so-called single fluid model. For the equation system to be complete it has to be supplemented by an *equation of state*, which describes the relation between pressure and density in the plasma. Depending on the physical context it is often sufficient to use either the isothermal equation of state,

$$p = nk_B T \qquad (3.12)$$

which holds for a constant temperature system, or the adiabatic equation of state

$$\frac{p}{n^\gamma} = constant \qquad (3.13)$$

where $\gamma = c_p/c_v = (2+N)/(N)$ is the adiabatic component defined by the specific heats at constant pressure (c_p) and volume (c_p) respectively or via the degrees of freedom of the system N_{free}. We will see in Section 3.8 that only for the stages in the ICF process where the density of the plasma is very high do more complex equations of state have to be used.

In ICF physics, Eqs. 3.9 and 3.10 can often be simplified to some extent. In this context in most cases the magnetic field is small enough in comparison to the electric field that it can be neglected, which leads to the following set of equations:

$$\frac{\partial \rho}{\partial t} + \nabla(\rho \mathbf{v}) = 0,$$

$$\rho \left[\frac{\partial \mathbf{v}}{\partial t} + (\mathbf{v} \cdot \nabla)\mathbf{v} \right] = -\nabla P + \frac{\rho}{m}\mathbf{F},$$

$$\frac{\partial \rho \epsilon}{\partial t} + \nabla \left[\mathbf{v}(\epsilon + P) + \kappa \nabla T - \Phi \right] = 0.$$

When $\epsilon = p/(y-1) + 1/2\rho v^2$ is inserted into the general equation of motion, the so-called Navier-Stokes equations are obtained as

$$\frac{\partial \rho}{\partial t} + \nabla(\rho \mathbf{v}) = 0, \qquad (3.14)$$

where μ is the shear viscosity and κ is the thermal conductivity. This equation system can now be used to describe the dynamical development of the plasma.

As already mentioned, the one-component description is only valid as long as no charge separation occurs over the length scales of concern. However, in some cases the plasma has to be modeled as a two-component system, taking care of the electrons and ions separately. This means the complete plasma description requires six instead of three equations, with the added complication of considering the interaction between the electrons and ions via the Maxwell equations. Such an extended set of magnetohydrodynamic equations can be quite complex.

Fortunately, it is often not necessary to apply the two-component description to its full extent. If the Debye length is much shorter than the mean free path,

$$\lambda_D \ll \lambda_{ee}, \tag{3.17}$$

where the mean free path is given by $\lambda_{ee} = 1/n\sigma$, and σ is the collisional cross-section. Relation 3.17 means implicitly that the charge density and velocity of the electrons and ions can be assumed to be equal

$$n_e \sim Z n_i,$$
$$v_e \sim v_i.$$

In this case introducing separate temperatures for the electrons and ions, T_e and T_i, respectively with

$$T_e \neq T_i$$

a description equivalent to Eqs. 3.14–3.16 suffices. This is valid if the timescale τ of the processes of interest is shorter than the electron-ion temperature equilibration time τ_{ei},

$$\tau \ll \tau_{ei}. \tag{3.18}$$

Typically inertial fusion plasmas can be described as a single fluid with two temperatures. In practice this is done as follows:

Because the mass of the ions is much larger than that of the electrons, the ions are responsible for the momentum transport and the electrons for the energy transport. The momentum transport is directly coupled to the

$$\frac{\partial \rho}{\partial t} + \nabla(\rho \mathbf{v}) = 0$$

$$\rho \left(\frac{\partial}{\partial t} + \mathbf{v}\nabla \right) \mathbf{v} = -\nabla p + \frac{1}{3}\nabla \mu \nabla \mathbf{v} + \frac{\rho}{m}\mathbf{F}$$

$$\rho c_{ve} \left(\frac{\partial}{\partial} + \mathbf{v}\nabla \right) T_e = \nabla \kappa_e \nabla T_e - p_e(\nabla \mathbf{v}) - \omega_e i(T_e - T_i) + S_e$$

$$\rho c_{vi} \left(\frac{\partial}{\partial} + \mathbf{v}\nabla \right) T_i = \nabla \kappa_i \nabla T_i - p_i(\nabla \mathbf{v}) + \omega_e i(T_e - T_i) + S_i \quad (3.19)$$

This system of equations is used to describe the dynamics in cases, in which the electron and ion temperature are only weakly coupled — for example, modeling of the corona and the ablation zone would need this kind of treatment.

By contrast, for investigations of the cold compressed plasma core it is sufficient to use a single temperature description, because the equilibration time between the electrons and ions is very short ($\sim 10-12$ ps) compared to the dynamical timescales ($\sim$ ns).

Again, analytical solutions of these equation systems exist for a very limited number of simple cases, so complex computer codes are normally needed to solve this set of equation systems.

3.4 Plasma Waves

An important property of plasmas is their ability to transmit collective disturbances or waves, which in the simplest case are just fluctuations in the electron or ion density. The understanding of wave phenomena is of special importance in the context of ICF as the laser light can interact not only directly with the plasma particles but also with the plasma waves (Kruer, 1988; Liu and Tripathi, 1995).

Here it should be noted that the plasma oscillations with frequency ω_p are **not** strictly waves (see Section 3.1) because they are stationary. An an essential aspect of waves, however, is the propagation of the electron or ion density fluctuations.

In the following the fluid description will be used to investigate the different types of waves that can be present in ICF plasmas. For the first type of wave — the Langmuir wave — it is assumed that the ions are at rest. This is equivalent to setting $T_i = 0$ and only the electrons have to be considered in the mass, momentum, and energy conservation equations, so

$$m_e n_e \left[\frac{\partial v_e}{\partial t} + v_e \frac{\partial v_e}{\partial x} \right] = -e n_e E - \frac{\partial}{\partial x} P_e$$

$$\frac{d}{dt} \left(\frac{P_e}{n_e^3} \right) = 0. \tag{3.20}$$

Combining the equation system with the Poisson equation for stationary ions,

$$\frac{\partial E}{\partial x} = -\frac{e}{\epsilon_0} (n_e - Z n_0),$$

where $n_i = n_0$ and $v_i = 0$, a solution can be found by assumes that the electron movement is small, allowing the fluid quantities to be linearized:

$$n_e = n_0 + n_1$$
$$v_e = v_1$$
$$p_e = n_0 kT + p_1$$
$$E = E_1.$$

In addition it is assumed that the adiabatic equation of state for an ideal gas given by Eq. (3.13) can be applied. Because $n_1 \ll n_0$, the nonlinear terms such as $n_1 v_i$, $v_1(\partial v_1 / \partial x)$ can be neglected and it follows that

$$\frac{\partial n_1}{\partial t} + n_0 \frac{\partial v_1}{\partial x} = 0$$

$$m n_0 \frac{\partial v_1}{\partial t} = -e n_0 E_1 - \frac{\partial P_1}{\partial x} \tag{3.21}$$

$$\frac{\partial E}{\partial x} = -\frac{e n_1}{\epsilon_0}$$

$$\frac{P_1}{P_0} = \gamma \frac{n_1}{n_0}.$$

From the last equation it follows that assuming an adiabatic equation of state (i.e. $P_0 = n_0 k_B T$ for the electrons), the background ions have automatically an isothermal equation of state $P_1 = 3k_B T_e n$. Elimination of E_1, P_1, and v_1 in the equation system 3.21 leads to the wave equation

$$\left(\frac{\partial^2}{\partial t^2} - \frac{3k_B T_e}{m} \frac{\partial}{\partial x^2} + \omega_p^2 \right) n_1 = 0. \tag{3.22}$$

Because we are interested in the harmonic solutions we apply the Ansatz

$$A = A_0 e^{i\omega t - kx}$$

which is characteristic for an electron plasma wave. This type of wave is also called Langmuir or plasma wave. The phase nelocity $v_\Phi = \omega/k$ and the group velocity $v_g = d\omega/dk$ are connected to the thermal velocity by the dispersion relation

$$v_g v_\phi = 6v_{th}^2.$$

Apart from Langmuir waves, the other types of plasma waves important in the ICF context are ion acoustic waves and electromagnetic waves. For all these waves dispersion relations can be derived in a similar fashion as demonstrated above. The corresponding relations for the ion acoustic wave and the electromagnetic wave are given by

$$\text{ion acoustic:} \qquad \omega^2 = k^2 \frac{Zk_B T_e + 3k_B T_i}{m_i} \qquad (3.24)$$

$$\text{electromagnetic:} \qquad \omega^2 = \omega_p^2 + c^2 k^2. \qquad (3.25)$$

For a detailed derivation of these dispersion relations the reader is referred to Kruer (1988). It should be noted that Langmuir waves and ion acoustic waves are longitudinal waves, whereas electromagnetic waves are of transversal character. Even more types of waves exist if strong magnetic fields are present in the plasma. However, they play a minor role in the context of ICF (Stix, 1992).

3.5 Plasma Heating

Returning to the electrostatic waves, we note that these waves cannot easily leave the plasma, because $v_g \sim \sqrt{3}v_t \ll c$. Therefore the energy of the waves will be dissipated to the plasma particles via some damping mechanism. There are actually several different types of damping at work, of which the most important are

• collisional damping or inverse bremsstrahlung and
• Landau damping.

Collisional Damping

Collisional or inverse bremsstrahlung damping is the process that converts the coherent movement of the electrons in the wave into thermal energy. The energy conversion proceeds through electron-ion collisions and will be

over the electron distribution, which is $_{kin}$ $_{ei}$ $_e$ $_e/2$, where $_{ei}$ the electron-ion collision frequency and v_e the electron velocity gained from the field given by the Lorentzian equation of motion as $v_e = eE/m\omega$. From $E_{field} = E_{kin}$ it follows

$$E_{field} = \frac{\epsilon_0 \nu E^2}{2} = \nu_{ei} \frac{n_e m v_e^2}{2} = E_{kin}.$$

The damping rate of inverse Bremsstrahlung is therefore given by

$$\nu = \frac{\omega_p^2}{\omega^2} \nu_{ei}. \tag{3.26}$$

Landau Damping

Landau damping can occur even in the absence of collisions (i.e., in the limit $\nu_{ei} \to 0$). This is the case for a plasma with a very high temperature T_e. In a simple one-dimensional picture, the motion of the electrons and ions in the plasma can be described in the presence of an electrostatic wave $E \sin(kx - \omega t)$ by

$$\frac{dv}{dt} = \frac{qE}{m} \sin(kx - \omega t). \tag{3.27}$$

Most particles will have a velocity either much smaller, $v \ll \omega/k$ or higher $v \gg \omega/k$ than that of the wave. These particles oscillate in the field without any energy change.

However some particles will have a velocity $v \sim \omega/k$ close to the wave's velocity, these particles experience a nearly constant field strength and are either slowed down or accelerated. If the velocity distribution is not constant, the particles with a velocity close to the wave velocity can become resonant with the field. Using a linearized version of the Vlasov Eq. 3.4, one can calculate the damping rate caused by these resonant particles. The wave damping rate γ_L because of this so-called Landau damping is given by

$$\frac{\gamma_L}{\omega} = -\sqrt{\frac{\pi}{8}} \frac{\omega_p^2 \omega}{k^3 v_t^3} \exp\left(-\omega^2/2k^2 v_t^2\right). \tag{3.28}$$

For a detailed calculation derivation of this damping mechanism see Chen (1984).

$$E_y = E_0 \sin(\omega t - kx),$$
$$B_z = \frac{E_0}{c} \sin(\omega t - kx).$$

The motion to the second order in field amplitude is given by

$$y = \frac{-v_{osc}}{\omega} \sin(\omega t - kx),$$
$$x = \frac{-v_{osc}^2}{4\omega c} \left(\frac{\cos 2(\omega t - kx)}{2} - \omega t + kx \right),$$

where $v_{osc} = eE/m\omega$ is the electron "quiver" velocity. In a homogeneous field the time-averaged force vanishes. This is not so in the case of an inhomogeneous field, where the force is determined by the gradient of the field intensity,

$$\mathbf{f}_{pond} = -\frac{1}{4} \frac{e^2}{\omega^2 m} \nabla E_0^2.$$

Ponderomotive forces have to be taken into account if $v_{os} > v_t$. This condition can be rewritten as a relation for $I\lambda^2$. In the case of an inhomogenous plasma with temperature in the range of 100–1000 eV, the equivalent ponderomotive pressure $P_L = \epsilon_0 E_0^2/2c$ becomes important if it exceeds the local thermal pressure $P_e = n_e k_B T_e$, or if

$$I\lambda_{\mu m}^2 \gtrsim 5 \times 10^{15} T \text{ [keV Wcm}^{-2}\mu m^2]. \tag{3.29}$$

3.7   Shock Waves

In the introduction it was pointed out that shocks play a dominant role in the compression phase of ICF. Here we discuss the basics of shock development and their propagation in plasmas more quantitatively.

If a plasma is disturbed very strongly in an extremely short time, as in the case of energy deposition by the driver, the disturbance propagates into the neighboring plasma regions approximately with the speed of sound c_s. Because the sound speed is proportional to the square root of the plasma density,

$$c_s \sim \rho^{1/2}, \tag{3.30}$$

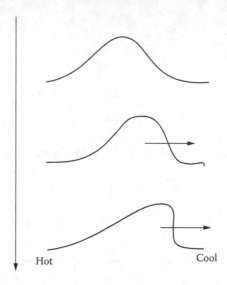

Hot

Cool

Figure 3.3. Temporal development of a shock structure in a plasma.

disturbances propagate faster in high-density regions than in low-density regions. Therefore, if a fast propagating disturbance travels into a lower density region, the perturbation profile will steepen, resulting in a shock wave, as illustrated in Fig. 3.3. This shock wave is supersonic, propagating faster than the sound speed of the lower density plasma lying ahead.

The so-called Mach number M is commonly used to characterize the shock wave and is defined by the ratio of the shock velocity v_{shock} to the sound speed of the plasma ahead of the shock front $v_{s,0}$,

$$M = \frac{v_{shock}}{v_{s,0}} > 1. \qquad (3.31)$$

Because the shock velocity is higher than the sound speed, the Mach number is always greater than 1.

Mathematically such a shock can in principle be described as a discontinuity in density, velocity, and temperature in the hydrodynamic equations. In reality the shock front is not infinitesimally small, because viscosity and thermal conductivity effects tend to smear it out somewhat.

In the following, the shock propagation in the medium is described for the simple case of a planar shock as illustrated in Fig. 3.4. The plasma behind and ahead of the shock are assumed to be in steady state. The plasma ahead of the shock described by p_0 and ρ_0 is assumed to be at

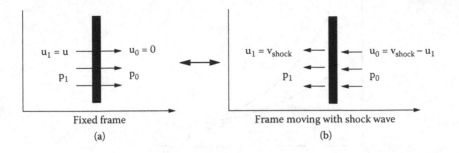

Figure 3.4. Frames of reference for shock description.

rest (i.e., $u_0 = 0$); behind the shock (p_1, ρ_1), the plasma moves initially with the velocity u_1. It is convenient to shift to a frame comoving with the shock wave frame (Fig. 3.4b). The conservation laws are then given in one-dimensional form by

$$\frac{\partial \rho}{\partial t} + \nabla(\rho u) = 0,$$

$$\frac{\partial}{\partial t}(\rho u) + \frac{\partial}{\partial x}(p + \rho u^2) = 0,$$

$$\frac{\partial}{\partial t}\left(\rho c_v kT + \frac{\rho u^2}{2}\right) + \frac{\partial}{\partial x}\left[\rho u\left(c_v kT + \frac{u^2}{2} + \frac{p}{\rho}\right)\right] = 0. \quad (3.32)$$

In a steady state the time derivatives can be ignored. In this case the system of equations (3.32) simplifies to

$$\rho_0 u_0 = \rho_1 u_1$$

$$p_0 + \rho_0 u_0^2 = p_1 + \rho_1 u_1^2$$

$$c_v kT + \frac{u_0^2}{2} + \frac{p_0}{\rho_0} = c_v kT + \frac{u_1^2}{2} + \frac{p_1}{\rho_1}.$$

These equations are known as the Ranking–Hugoniot relations. Using the definitions $V_0 = 1/\rho_0$ and $V_1 = 1/\rho_1$ of the specific volumes V_0 and V_1, we arrive at the relations

$$u_0^2 = V_0^2\left(\frac{p_1 - p_0}{V_0 - V_1}\right)$$

$$u_1^2 = V_1^2\left(\frac{p_1 - p_0}{V_0 - V_1}\right). \quad (3.33)$$

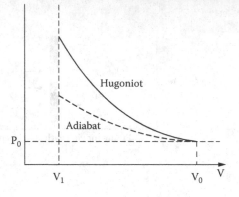

Figure 3.5. p-V diagram of Hugoniot and adiabatic curve.

Also, because

$$\frac{1}{2}(u_0 - u_1) = \frac{1}{2}(p_1 - p_0)(V_0 - V_1)$$

it follows that

$$c_v k(T_1 - T_0) = \frac{1}{2}(p_1 - p_0)(V_0 - V_1).$$

Combining this equation with the equation of state of the plasma, one finds that the pressure behind the shock depends only on the pressure in front of the shock and the densities in front and behind,

$$p_1 = G(p_0, \rho_0, \rho_1). \tag{3.34}$$

This relation is known as the shock Hugoniot relation. As illustrated in the p-V diagram of Fig. 3.5, the Hugoniot curve lies above that of adiabatic compression. The area under the curve is equivalent to the work done to compress the plasma, so we see immediately that it requires more work to shock-compress the plasma than it does adiabatically.

If we assume the plasma to be an ideal gas, it follows that

$$c_v T = \frac{1}{\gamma - 1} pV$$

$$c_p T = \frac{\gamma}{\gamma - 1} pV. \tag{3.35}$$

After a short calculation the relationship between the upstream and downstream variables is then given by

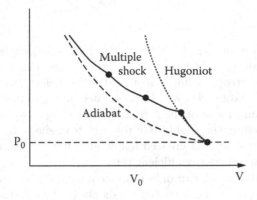

Figure 3.6. Comparison between Hugoniot shock and multiple shock compression.

$$\frac{\rho_1}{\rho_0} = \frac{(\gamma+1)p_1 - (\gamma-1)p_0}{(\gamma-1)p_1 - (\gamma+1)p_0}$$

$$\frac{p_1}{p_0} = \frac{(\gamma+1)\rho_1 - (\gamma-1)\rho_0}{(\gamma-1)\rho_0 - (\gamma+1)\rho_1}$$

$$\frac{T_1}{T_0} = 1 + \frac{2\gamma}{(\gamma+1)^2}\frac{\gamma M_0^2 + 1}{M_0^2}(M_0^2 - 1).$$

Substituting the specific volume into the pressure equation, one obtains the better known explicit form of the shock Hugoniot relation of an ideal gas:

$$p_1 = p_0 \frac{(\gamma+1)V_0 - (\gamma-1)V_1}{(\gamma+1)V_1 - (\gamma-1)V_0} = H(V_1, p_0, V_0). \qquad (3.36)$$

In the limit of very weak shocks the propagation of a disturbance approaches that of an isentropic sound wave. In ICF, the opposite case of very strong shocks is much more important. A strong shock means $p_1 \longrightarrow \infty$ and it follows for the densities

$$\frac{\rho_1}{\rho_0} = \frac{(\gamma+1)(p_1/p_0) - (\gamma-1)(p_0/p_0)}{(\gamma-1)(p_1/p_1) - (\gamma+1)(p_0/p_1)} \longrightarrow \frac{\gamma+1}{\gamma-1}.$$

For a monoatomic gas (i.e., $\gamma = 5/3$), it follows that

$$\frac{\rho_1}{\rho_0}(\text{monoatomic gas}) \longrightarrow \frac{\gamma+1}{\gamma-1} = 4, \qquad (3.37)$$

which means that the maximum compression even by an infinitely strong single shock can never produce more than four times the initial density. In

one. This has two advantages. First of all, it is more energy efficient to use a series of weak shocks, as illustrated by Fig. 3.6: one can stay closer to an adiabatic and isentropic compression and nevertheless reach the same pressure as a single strong shock. Moreover, it is indeed more efficient to use weaker shocks and more of them (illustrated in Fig. 3.6) than a single strong shock. Therefore, the goal in ICF research is to shape the laser pulse in such a way that the shocks created follow an adiabatic compression as closely as possible in an energy efficient way.

Although a single shock can only produce a compression of a factor 4, this enhancment applies to each successive shocks in turn, so that n shocks can result in a compression of a factor 4^n, resulting a much higher total compression.

The above picture oversimplifies the situation to some extent, because the shocks were treated assuming the plasma to behave like a single component ideal gas. However, the plasma is really a two-component, nonideal system leading to much more complex shock structures (Gross and Chu, 1969). To start with, the shock is assumed to propagate in an ideal gas, in which case the shock is can be described as a sharp disconuity in density. For a nonideal gas, a shock has a certain thickness which is typically of the order of several mean-free paths of the gas particles (see Fig. 3.7a) resulting from the gas viscosity. In a plasma, thermal conductivity is determined by the electrons, whereas viscosity is due to the ions. The shock density profile is mainly determined by the ions, so that the shock thickness is of the order of the mean free path of the ions.

However, the electrons influence the shock profile as well. In Section 3.3 we saw that an ICF plasma can often be described by a single fluid, two-temperature model. The driver energy is deposited in the electrons, increasing the electron temperature behind the shock. However, the high thermal conduction of the electrons allows the electrons to transport this thermal energy ahead of the shock, where it is then transferred to the ions via electron-ion collisions. In this way ions are preheated in front of the shock wave. The shock structure in the plasma has therefore the overall form illustrated in Fig. 3.7b.

The existence of the preheat "foot" ahead of the shock reduces the shock strength and therefore the compression of the plasma shock wave. This effect is obviously highly undesirable, as it can considerably reduce the efficiency of the compression. Avoiding preheat is therefore a major consideration when creating a succession of shocks.

The above description of shocks is still not sufficient to understand compression in the ICF context, in which one has to move beyond the model

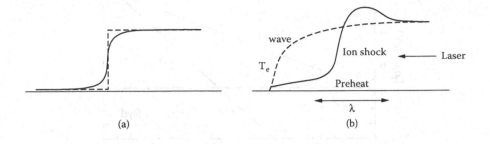

Figure 3.7. Shock profiles in a) real gas and b) plasma.

of planar shocks to that of spherical shocks. The spherical geometry is the key to achieving the required densities of $\sim 10^3$ g/cm^3 for fusion (planar shocks could never reach these densities). These spherically symmetric shocks will be investigate in detail later in Chapter 5 while we continue here with our introduction of the basic underlying plasma physics. Next we turn to the question of which equation of state is needed to correctly describe an ICF plasma.

3.8 Equation of State for Dense Plasmas

In the previous sections it was assumed that the plasma behaves more or less like an ideal gas; the only modifications to this picture came through the fact that the plasma consists of two components — namely electrons and ions. Therefore the equation of state (EOS) for an ideal gas has the form,

$$p = nk_BT = \rho RT, \tag{3.38}$$

where R is the gas constant per mass unit, or as in Eq. 3.12 was used. The application of the EOS of an ideal gas is a valid theoretical treatment for many stages of the fusion process, but not for all of them. Proceeding through the different stages of the ICF process, we consider which modifications to the EOS are necessary.

When the driver hits the target, the plasma is heated; here, it is important to include the effects of excitation and ionization. Ionization begins for most materials at 7–15 eV, well below the actual ionization potential. The internal energy is modified in such a way as to include the changes in potential and thermal energy of the electrons and ions through ionization. As the temperatures rise higher the radiation pressure can become comparable to the hydrodynamic pressure of the gas. If there exists thermal

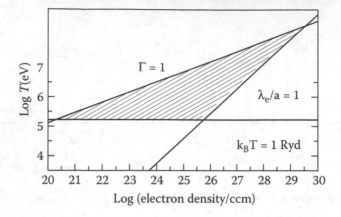

Figure 3.8. Temperature and density regimes where electron degeneracy and coupling effects play an important role in the equation of state for the example of a fully ionized hydrogen plasma.

equilibrium with the gas, the radiation pressure $p_{rad} = \sigma T^4$ ($\sim$ black body radiation) can simply be included in the EOS,

$$p = nk_B T + p_{rad}.$$

However, we saw that a central requirement for ICF to work is to achieve very high temperatures and densities in the central region. A plasma at such high density will differ considerably from an ideal gas, because two effects become important, namely coupling effects and electron degeneracy.

Figure 3.8 illustrates the different regions in the density-temperature diagram where these effects are important. For example, at high densities and low electron temperatures, electron degeneracy dominates, whereas for high temperatures and high densities, coupling effects play a more important role. In the following only the principles of the much more complex equation of states of high density plasmas will be described — see Eliezer *et al.* (2002) for more details.

Strongly Coupled Plasma

An important parameter which characterizes the strength of the interaction between the particles is the ion coupling parameter Γ. It essentially describes the ratio of the potential to the kinetic energy of the particles of

is the charge of the particles, $= (4\pi n_i/3)$ is the interparticle spacing, and n_i is the ion density. In plasmas with $\Gamma \gtrsim 1$ the potential energy dominates over the kinetic energy and the plasma is strongly coupled. Outside the ICF context, such strongly coupled plasmas occur naturally in planetary (Stevenson, 1982) and stellar (Chaouacha *et al.*, 2004) interiors, in white dwarfs (Chabrier *et al.*, 1992) and in the crust of neutron stars. Experimentally, they can be produced by high-intensity, short-pulse lasers, too (Perry and Mourou, 1994).

Such plasmas have densities equivalent to that of solids, but temperatures are far higher, typically in the range of 1 eV–50 keV. The strong Coulomb coupling significantly affects the basic properties of the system, not only the EOS but also transport processes. The reason for this is that, at high densities, collisions become increasingly important. It is not only the number of collisions but also that large-angle scatterings and three-body collisions are no longer rare events. All theories based on the assumption of small-angle pair-scatterings are therefore no longer applicable and the whole problem becomes strongly nonlinear.

Degenerate Electron Gas

For very high densities and relatively low temperatures (which can still be of the order of 100 000 K) quantum mechanical effects tend to dominate the behavior of the electron gas. In particular, the exclusion principle becomes important as the de Broglie wavelength $\lambda_{deBroglie}$ of the electrons is in certain temperature-density regimes comparable to the interparticle spacing a,

$$\lambda_{deBroglie} = \frac{\hbar}{\sqrt{2\pi m_{ij} k_B T}} \approx a = \left(\frac{3}{4\pi n_i}\right)^{1/3}. \tag{3.40}$$

Because the electrons are compressed beyond the deBroglie wavelength, only a limited number of quantum states are available.

If one of the energy levels is filled, no more electrons can be added to this level, but have to be added to a higher one. The electrons behave like a degenerate electron gas, where they no longer follow a Maxwell–Boltzmann distribution

$$n(\epsilon) \sim \exp[-\epsilon/kT]$$

but instead Fermi–Dirac statistics

$$n(\epsilon) \sim \frac{1}{\exp[(\epsilon - \mu)/kT] + 1}, \tag{3.41}$$

| Thermal conductivity | κ | $\sim T^{5/2}$ | $T T_F^{3/2}$ |

where ϵ is the electron energy. In case of a Fermi–Dirac distributed electron gas, any quantity that is obtained by averaging over the electron distribution is significantly altered from that of a Maxwell-Boltzmann electron gas. Examples of altered dependencies are given in Table 3.1 (Rose, 1988).

The energy of the maximum filled level ϵ_F — the so-called Fermi energy — is given by

$$\epsilon_F = \frac{1}{8}\frac{h^2}{m_e}\left(\frac{3n_e}{\pi}\right)^{2/3} = 2.19 \cdot 10^{-15} n_e^{2/3} \quad [\text{eV}] \tag{3.42}$$

and the the so-called Fermi temperature by

$$k_B T_F = \frac{\hbar}{2m}\left(3\pi^2 n_e\right)^{2/3} \sim 7.86\text{eV}\left(\frac{n_e}{10^{23}\text{cm}^{-3}}\right)^{2/3}. \tag{3.43}$$

The electron density and the maximum filled momentum state $p_F = \sqrt{2m_e\epsilon}$ are correlated via

$$n_e = \int_0^{p_F} \frac{8\pi p^2}{h^3}dp. \tag{3.44}$$

The usual condition for Fermi degeneracy is expressed by comparing the thermal energy with the Fermi energy, defining the so-called the electron degeneracy parameter. This is given by

$$\theta_e = \frac{k_B T}{\epsilon_F}. \tag{3.45}$$

If $\theta_e < 1$, the electron gas is fully degenerate. This condition of complete degeneracy is usually not fulfilled for ICF plasmas — whereas it is for electrons in a metal, where at solid densities the temperature is below the Fermi energy of 5 eV.

However, the distinction between degenerate and nondegenerate electron gas becomes less abrupt for higher temperatures. In this in between region the electron gas is called partially degenerate, for which the reader is referred to textbooks on statistical physics. In the context of ICF, the plasma in the highly compressed core will generally be partially degenerate until it ignites and heats to much higher temperatures. This of course complicates the theoretical treatment of the phase just before ignition.

be used if the plasma has highly degenerate electrons, F, and the plasma density is so high that $a < (\pi/12)^{2/3}(\hbar/me^2)$.

In general, the EOS of strongly coupled or degenerate plasmas is fairly complex and not completely understood, especially for heavier elements. A simple analytical expression such as Eq. 3.39 for such plasma does not exist. Analytical calculations mostly have to apply simplifying assumptions to treat such systems. One alternative is to simulate the microscopic behaviour of dense plasmas either by Monte Carlo methods (Hansen *et al.*, 1979) or molecular dynamics simulations (Pfalzner and Gibbon, 1996). In Monte Carlo methods one obtains the properties of the dense plasma from an ensemble of particle configurations weighted by a canonical ensemble distribution. In particle simulations the coupled equations of motion of interacting particles are integrated (ab initio). By averaging the microscopic trajectories in space and time the macroscopic properties can be deduced.

Alternatively, empirical information about the EOS can be obtained from high-density experiments. The EOS is usually given in tabulated form combining the information of analytical, numerical and experimental investigations. One often-used source for the equation of state are the so-called SESAME tables, see http://www.t4.lanl.gov/opacity/sesame.html.

Remark

This chapter is intended only as a short introduction of the plasma physics relevant to the ICF context. This is a very compressed overview and in no way a complete or detailed description. For a derivation of all the equations introduced here, the reader is refer to the above mentioned books.

In the following chapters we consider how one can understand the relevant processes in ICF by applying simple models, thereby enabling the beginner to understand the basic physical processes in ICF. However, for many physical phenomena in ICF these simple models and even more complex analytical description do not suffice to describe the situation realistically. In these cases numerical modeling is often the only way to investigate to investigate the physical processes theoretically. Because these codes are often very specialized and complex, it goes beyond the scope of this book to describe these numerical methods in detail at this point. However, at the end of each chapter it will be indicated which additional physical aspects are included when using such codes.

Chapter 4

Absorption of Laser Light

In the previous chapter most of the basic plasma phenomena needed to understand ICF were introduced. In this chapter we will look at which of these processes take place in the various stages of inertial confinement fusion. Starting from the moment the laser hits the target and a plasma created, we will see that the subsequent absorption of laser light is rather complex and many different processes take place simultaneously. As far as possible simple models will be used to describe the most important processes of this phase. In general sophisticated computer simulations are necessary to understand absorption physics quantitatively. At the appropriate places we will refer to further reading of these detailed studies.

4.1 Coupling of the Laser Energy to the Target

When the laser light hits the target, material immediately evaporates from the surface and a plasma layer is created. The whole region dominated by

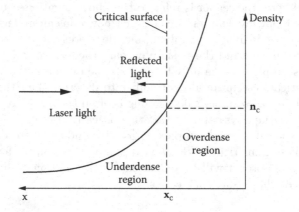

Figure 4.1. Formation of critical density surface.

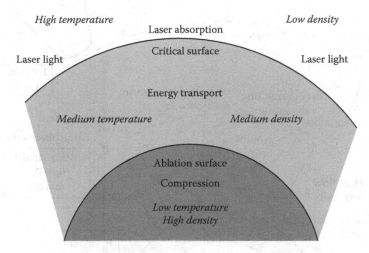

Figure 4.2. Schematic picture of laser plasma interactions.

laser-plasma interactions is called the corona. As mentioned in Chapter 2, the laser light can only penetrate the plasma as long as the plasma density is not higher than the critical electron density n_c given by

$$n_c = \frac{\epsilon_0 m \omega_L^2}{e^2} = 1.1 \times 10^{21} \left(\frac{\lambda_L}{1\mu m}\right)^{-2} \; [\text{cm}^{-3}], \qquad (4.1)$$

where ω_L and λ_L are the laser frequency and wavelength, respectively. If plasma regions are formed with a density higher than the critical density n_c, the plasma frequency becomes larger than the laser frequency (i.e., $\omega_p > \omega_L$). The density profile of the plasma ablated from the target surface is similar to Fig. 4.1. Most light is absorbed at or near the critical surface, where $n_e = n_c$.

As the interaction of the laser with the target continues, distinct regions develop — an absorption domain, a transport domain, and a compression domain, as illustrated in Fig. 4.2. These domains differ significantly in temperature and density.

The absorption domain is separated from the transport domain by the critical surface. The plasma here has a high temperature ($\sim$1000 eV) but a relatively low density of less than 0.01 g/cm^3. In the corona the electrons absorb the laser energy. This energy is then transported from the critical surface to the ablation surface, where the plasma is created. In this transport domain the density is between 0.01 g/cm^3 and solid density (0.2

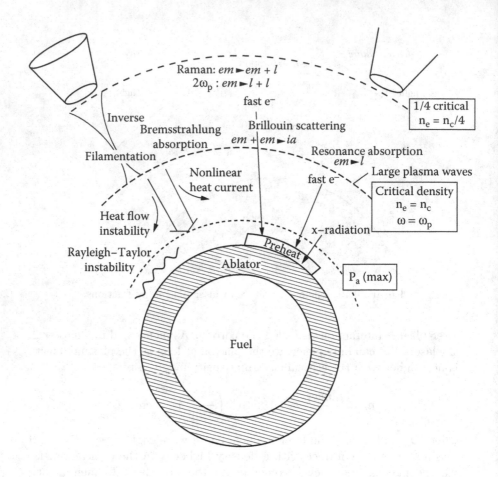

Figure 4.3. The different physical processes going on in the corona of an irradiated microballoon. em denotes electromagnetic waves, l Langmuir waves, and ia ion acoustic waves.

g/cm^3) with temperatures ranging between $\sim$30 eV and 1000 eV. Finally, in the compression domain, the densities range between solid density ρ_0 and 10 ρ_0 at temperatures of 1–30 eV.

At the ablation surface the plasma blows off in direction of the laser with approximately sound speed c_s. Figure 4.3 illustrates the multitude of physical processes going on in the corona. These processes will now be considered in detail.

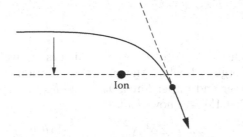

Figure 4.4. Coulomb scattering between an electron and ion. b denotes the impact parameter and θ the scattering angle.

4.2 Inverse Bremsstrahlung Absorption

Inverse bremsstrahlung absorption is an essential mechanism for coupling laser energy to the plasma. Laser light is absorbed near the critical surface via inverse bremsstrahlung the following way:

The electrical field induced by the laser the causes electrons in the plasma to oscillate. This oscillation energy is converted into thermal energy via electron-ion collisions, a process known as inverse bremsstrahlung. Bremsstrahlung and inverse bremsstrahlung are connected the following way: if two charged particles undergo a Coulomb collision they emit radiation — so-called bremsstrahlung. Inverse bremsstrahlung is the opposite process, where an electron scattered in the field of an ion absorbs a photon.

Using the notation as given by Fig. 4.4, the differential cross-section $d\sigma_{ei}/d\Omega$ for such a Coulomb collision is described by the Rutherford formula given by

$$\frac{d\sigma_{ei}}{d\Omega} = \frac{1}{4}\left(\frac{Ze^2}{m_e v^2}\right)^2 \frac{1}{\sin^4(\theta/2)}, \tag{4.2}$$

where θ is the scattering angle and Ω the differential solid angle, which in spherical coordinates with azimuthal symmetry are connected by

$$d\Omega = 2\pi \sin\theta d\theta. \tag{4.3}$$

The impact parameter b is related to the scattering angle via

$$\tan\frac{\theta}{2} = \frac{Ze^2}{m_e v^2 b}.$$

The total cross-section σ_{ei} for electron-ion collisions is then obtained by integration over all possible scattering angles. Using Eqs. 4.2 and 4.3 this

The integral from $\theta \longrightarrow 0$ to $\theta \longrightarrow \pi$, which is equivalent to $b \longrightarrow \infty$ and $b \longrightarrow 0$ diverges. Luckily the physical situation in a plasma allow us to define a lower and upper limit, b_{min} and b_{max}, respectively, for this integration, so that Eq. 4.4 now reads

$$\sigma_{ei} = \frac{\pi}{2} \left(\frac{Ze^2}{m_e v^2} \right)^2 \int_{b_{min}}^{b_{max}} \frac{\sin\theta}{\sin^4(\theta/2)} d\theta. \qquad (4.5)$$

The physical reason for the upper limit is Debye shielding (Sec. 3.1), which makes distant collisions ineffective. Therefore, in a plasma the b_{max} limit can be replaced by the Debye length λ_D. The lower limit b_{min} is often set equal to the de Broglie wavelength; however, Lifshitz and Pitaevskii (1981) showed that this approach as such is inadequate and derived the lower limit to be $b_{min} = Ze^2/k_B T_e$. The total cross-section in a plasma is then given by

$$\sigma_{ei} = \frac{\pi}{2} \left(\frac{Ze^2}{m_e v^2} \right)^2 \int_{Ze^2/k_B T_e}^{\lambda_D} \frac{\sin\theta}{\sin^4(\theta/2)} d\theta. \qquad (4.6)$$

Knowing the cross-section one can calculate the collision frequency in the plasma. The collision frequency ν_{ei} is defined as the number of collisions an electron undergoes with the background ions per unit time, and depends on the ion density n_i, the cross-section σ_{ei} and the electron velocity v_e

$$\nu_{ei} = n_i \sigma_{ei} v_e. \qquad (4.7)$$

In calculating the collision frequency for electron-ion collisions, one has to take the velocity distribution of the particles into account. In many cases it can be assumed that the ions are at rest ($T_i = 0$) and electrons are in local thermal equilibrium. A Maxwellian electron velocity distribution $f(v_e)$ (i.e., of the form)

$$f(v_e) = \frac{1}{(2\pi k_B T_e/m)^{3/2}} \exp\left[-\left(\frac{m_e v_e^2}{2k_B T_e} \right) \right] \qquad (4.8)$$

is isotropic and normalized in a way that

$$\int_0^\infty \frac{4\pi v_e^2}{(2\pi k_B T_e/m)^{3/2}} \exp\left[-\left(\frac{m_e v_e^2}{2k_B T_e} \right) \right] = 1.$$

Using Eqs. 4.4 and 4.8 and performing the integrations, the electron-ion collision frequency becomes

$$\nu_{ei} = \left(\frac{2\pi}{m_e} \right)^{1/2} \frac{4Z^2 e^4 n_i}{3(k_B T_e)^{3/2}} \ln\Lambda, \qquad (4.9)$$

In deriving this equation it was assumed that small angle scattering events dominate, which is a valid assumption if the plasmas density is not too high. For dense and for cold plasmas Eq. 4.9 is not really applicable because large angle deflections become increasingly likely, violating the small angle scattering assumption. If one uses the above method, the values of b_{min} and b_{max} can become comparable so that $\ln \Lambda$ eventually turns negative — an obviously unphysical result. In practical calculations a lower limit of $\ln \Lambda = 2$ is often assumed. However, for dense plasmas a more complex treatment should really be applied (Bornath et al., 2001; Pfalzner and Gibbon, 1998). Caution is also necessary if the laser intensity is very high, as in this case strong deviations from the Maxwell distribution can occur.

Returning to less dense plasmas, the question is: how is laser energy now absorbed in the plasma? In a simple model one can assume that the equation of motion of the electrons is governed by the electric field and the collisions and therefore described by

$$\frac{d\mathbf{v}}{dt} = \frac{e\mathbf{E}}{m_e} - \nu_{ei}\mathbf{v}.$$ (4.10)

Solving this equation together with the Maxwell equation for a monochromatic wave with frequency ω_L and a wave number k, one obtains

$$(\mathbf{k} \cdot \mathbf{E})\mathbf{k} - \left[k^2 - \frac{\omega_L^2}{c^2} + \frac{\omega_p^2 \omega_L^2}{c^2(\omega_L + i\nu_{ei})} \right] \mathbf{E} = 0.$$

The transverse waves couple directly to the laser field in vacuum and their dispersion relation is given by

$$c^2 k^2 = \omega_L^2 - \frac{\omega_p^2 \omega_L}{\omega_L + i\nu_{ei}}.$$ (4.11)

In the plasma corona the collision frequency is much smaller than the laser frequency, $\nu_{ei} \ll \omega_L$, and a Taylor expansion of Eq. 4.11 in ν_{ei}/ω can be performed. Retaining only the first three terms of the expansion yields

$$k^2 = \frac{\omega_L^2}{c^2} \left(1 - \frac{\omega_p^2}{\omega_L^2} + \frac{i\nu_{ei}\omega_p^2}{\omega_L^3} \right).$$

Solving this equation by expanding the root for $\nu_{ei} \ll \omega_L$ and $\omega_L^2 - \omega_p^2 \gg (\nu_{ei}/\omega_L)\omega_p^2$ gives

$$k = \pm \frac{\omega_L}{c} \left(1 - \frac{\omega_p^2}{\omega_L^2} \right)^{1/2} \left[1 + i\frac{\nu_{ei}}{2\omega_L} \frac{\omega_p^2}{\omega_L^2} \frac{1}{1 - \omega_p^2/\omega_L^2} \right].$$

$$\kappa_{ib} = 2\Im(k) = \frac{\nu_{ei}}{c}\frac{\omega_p}{\omega_L^2}\left(1 - \frac{\omega_p}{\omega_L^2}\right) \qquad . \tag{4.12}$$

Using the definition of the critical density (Eq. 4.1), one obtains

$$\kappa_{ib} = \frac{\nu_{ei}(n_c)}{c}\frac{n_e^2}{n_c^2}\left(1 - \frac{n_e}{n_c}\right)^{-1/2} \tag{4.13}$$

and substituting the expression for the collision frequency (Eq. 4.9)

$$\kappa_{ib} \sim \frac{Z_i}{T_e^{3/2}}n_e^2\left(1 - \frac{n_e}{n_c}\right)^{-1/2}. \tag{4.14}$$

The dependence of κ_{ib} on n_e/n_c reflects the fact that a large fraction of the inverse bremsstrahlung absorption occurs near the critical density.

If laser light is absorbed, it means that the laser intensity I changes while it passes through a plasma. If we assume that the laser light moves in z-direction, the actual change of the laser intensity I in the plasma is described by

$$\frac{dI}{dz} = -\kappa_{ib}I. \tag{4.15}$$

In general the solution of Eq. 4.15 is complicated by the fact that the plasma is usually inhomogeneous. An additional difficulty arises from the fact that κ_{ib} depends on the electron density, temperature and the Coulomb logarithm and all these values are time-dependent.

However, for laser pulses of 1 ns duration or longer with intermediate intensities ($<10^{14}$ W/cm^2), the electron temperature and Coulomb logarithm can be assumed to constant. In this situation Eq. 4.15 can be solved analytically. The absorption coefficient over a length l is related to the difference between the incoming I_{in} and outgoing I_{out} laser intensity by

$$\alpha_{ib} = \frac{I_{in} - I_{out}}{I_{in}} = 1 - \exp\left[-\int_0^l \kappa_{ib}dz\right]. \tag{4.16}$$

For a linear density profile of the form $n_e = n_c(1 - z/l)$, Ginzburg (1961) showed that the analytical solution is given by

$$\alpha_{ib} = 1 - \exp\left[-\frac{32}{15}\frac{\nu_{ei}(n_c)l}{c}\right] \tag{4.17}$$

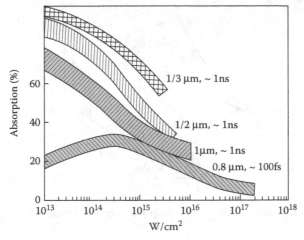

Figure 4.5. Experimental data of the laser absorption in solid low-Z targets (from Eliezer, 2002).

and Kruer (1988) demonstrated that for an exponential profile of the form $n_e = n_c \exp[-z/l]$, one obtains

$$\alpha_{ib} = 1 - \exp\left[-\frac{8}{3}\frac{\nu_{ei(n_c)}l}{c}\right]. \tag{4.18}$$

For more complex density profiles only numerical solutions exist.

A more detailed calculation of the inverse bremsstrahlung absorption requires the use of kinetic theory to take into account the electron distribution function and the position of the ions (Dawson, 1968). In this case the absorption coefficient depends on the ion correlation function. These effects are often described as nonlinear inverse bremsstrahlung, a detailed account of which can be found in Sid (2003).

Although inverse bremsstrahlung is an important absorption process in plasmas, it is not the only one. Inverse bremsstrahlung absorption is only efficient if enough collisions take place. However, from Eq. 4.9 it follows that the collision frequency ν_{ei} scales as $T_e^{-3/2}$. This means for higher temperatures, which also means higher laser energies, inverse bremsstrahlung becomes less and less effective. This effect has been verified in many experiments in which the absorption coefficient for a low-Z solid like aluminum was measured for different laser input energies. Figure 4.5 shows the result for different wavelength and pulse durations of the laser light.

However there are other processes that couple the energy of the laser light at higher laser intensities into the plasma; in particular, resonance absorption and parametric instabilities, which we consider in the following two sections.

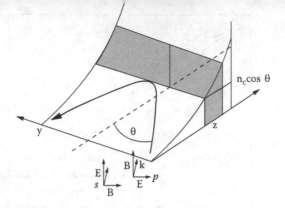

Figure 4.6. Schematic picture of p-polarization for resonance absorption.

4.3   Resonance Absorption

As illustrated in Fig. 4.1 the plasma that is created by the interaction of the laser with a solid target has an inhomogeneous density profile comprising both under- and overdense regions. Whenever light meets a plasma with these characteristics, electrostatic waves are excited if any component of the electric field of the light coincides with the direction of the density gradient (p-polarized interaction). In this case the electric field becomes very large near the critical surface, and it is here that waves are resonantly excited. In this way energy is transfered from the electromagnetic into plasma waves. Because these waves are damped, energy will eventually be converted into thermal energy, thus heating the plasma. This entire process of converting laser energy into plasma heating via wave excitation is called *resonance absorption* and can be schematically described by the following diagram:

One damping mechanism for these waves is obviously collisions (Section 4.2). However, there are also damping mechanisms of collisionless nature as for example Landau damping (Section 3.5) and wave breaking. A simple model for resonance absorption (Duderstadt and Moses, 1982; Kruer, 1988) shows how this happens: assuming a nonuniform plasma driven by a uniform electric field of strength E_u and frequency ω_0 and combining the Maxwell equations

$$\nabla \cdot \mathbf{E} = \frac{\rho}{\epsilon_0},$$

$$\frac{\partial \rho}{\partial t} + \nabla \cdot \mathbf{J} = 0,$$

This means that $\mathbf{J} + \epsilon_0 \partial \mathbf{E}/\partial t$ is a spatially independent component or

$$\frac{\partial \mathbf{E}}{\partial t} + \frac{\mathbf{J}}{\epsilon_0} = \left\langle \frac{\partial \mathbf{E}}{\partial t} + \frac{\mathbf{J}}{\epsilon_0} \right\rangle.$$

Neglecting the ion motion and assuming the oscillation is in z-direction, J depends on the oscillation velocity v_{osc} as $J = -en_0(z)v_{osc}$. Linearizing J, performing the temporal differentiation and using the linearized equation of motion (Eq. 4.10), it follows that

$$\frac{\partial^2 \mathbf{E}}{\partial t^2} + \omega_p^2(z)E + \nu_{ei}\frac{\partial \mathbf{E}}{\partial t} = -\left[\omega_p^2(z) - \langle \omega_p^2(z)\rangle\right] E_d \cos \omega_0 t.$$

Assuming an harmonic dependence of E (i.e., $E \sim \exp(i\omega t)$), the electric field response will be

$$E = \frac{\omega_p^2(z)E_d}{\omega_0^2(z) - \omega_p^2(z) + i\nu_{ei}\omega_0}. \tag{4.20}$$

The question of interest is, how much of the power is absorbed in the plasma? For a linear density gradient the absorbed power is given as

$$P_{ra} = \epsilon \int \frac{\nu_{ei}|E|^2}{2} d. \tag{4.21}$$

Substituting Eq. 4.20 one finally obtains

$$P_{ra} = \frac{\omega_0 l E_d^2}{8}. \tag{4.22}$$

Note that the dependence on the collision frequency in Eq. 4.21 cancels out: in other words, the resonance absorption process is independent of the details of the damping process. This means that, in contrast to inverse bremsstrahlung, resonance absorption can be efficient even for very low electron-ion collision frequencies. Therefore, resonance absorption can dominate over inverse bremsstrahlung absorption for high plasma temperatures, low critical densities, and short plasma scale-length. Put another way, resonance absorption is the main absorption process for high laser intensities and long wavelengths. More generally the laser light will be *obliquely* incident on the density gradient in the plasma. This situation is illustrated in Fig. 4.6. In this case the dispersion relation reads as

$$\omega_0^2 = \omega_p^2 + \omega_0^2 \sin^2\theta + k_z^2 c^2.$$

$$E_{||} \sim \frac{\sin \theta}{(\omega_0 l/c)^{1/16}} \exp \left[\frac{}{3}(\omega_0 l/c) \sin \theta \right], \qquad (4.23)$$

where l is the scaling height of the density gradient. This is the component that drives the resonant process and can be substituted in Eq. 4.20 for E_d.

Laser absorption is only possible if some kind of damping takes place, which can be either through electron-ion collisions or wave-particle interactions. Regarding the latter as a type of collision, an effective collision frequency ν_{eff} can be introduced, which includes both types of damping. The absorbed energy flux I_{ra} can then be presented as

$$I_{ra} = \frac{\epsilon_0}{2} \int \nu_{\text{eff}} E dz$$

integrating along the path of the electromagnetic ray. The fraction f_{ra} of energy absorbed through resonance absorption is then

$$f_{ra} = \frac{I_{ra}}{I_L} = \frac{\int \nu_{ei} E^2 dz}{cE_L^2}. \qquad (4.24)$$

Ginzburg (1961) and Pert (1978) showed that for a linear density profile the fraction of absorbed energy of a p-polarized wave is

$$f_{ra} \sim 36\tau^2 \frac{(\text{Ai}(\tau))^3}{(|d\text{Ai}(\tau)/d(\tau)|}, \qquad (4.25)$$

where Ai is the Airy function and

$$\tau = (\omega l/c)^{1/3} \sin \theta. \qquad (4.26)$$

The maximum of the fractional absorption lies at about $\tau \sim 0.8$, which, according to Eq. 4.26, corresponds to $\sin \theta = 0.42\lambda_L/l$ with a value of $f_{ra}(\tau \sim 0.8) \sim 0.5$. This means that for p-polarized laser light of $I_L\lambda_L^2 > 10^{15}$ (W/cm^2), up to 50% of the absorption can occur through resonance absorption.

This maximum absorption rate will be rarely achieved in an ICF scenario. The reason is that the range of angles θ depends on the scale-length l of the density gradient — the larger the scale-length the smaller the optimal angle of incidence. It would be difficult to combine this optimum angle of incidence with the requirement of illumination uniformity. In addition, ponderomotive forces (Section 3.6) because of the light pressure steepen the density profile and therefore shorten the scale length, so that resonance absorption is increased. This effect is of nonlinear nature and needs a more

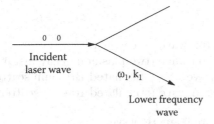

Figure 4.7. Schematic picture of three-wave parametric processes.

comprehensive numerical treatment than that which can be provided by simple theory.

Although resonance absorption can be very efficient where inverse bremsstrahlung absorption fails, there is a price to pay: the main feature of resonance absorption is that only a small fraction of the plasma electrons carry most of the absorbed energy. This means that as a side effect, a great many unwanted hot electrons are created, which preheat the plasma in front of the shock. We will discuss in Section 4.6 how these hot electrons influence the energy transport in a negative way.

4.4 Parametric Instabilities

As already mentioned, a hot plasma readily supports the excitation of waves. Waves can enhance absorption, which would be an desirable effect but can also reduce the absorption of the laser light. Here we want to look into these wave excitation processes in more detail.

Most instabilities caused by intense laser light can be described by the resonant decay of the incident wave into two new waves. This kind of process is called parametric instability. Although there also exist processes involving four or more waves, these play no significant role in the inertial confinement fusion (ICF) context. In general three-wave coupling can be represented by a schematic diagram like Fig. 4.7. Depending on which types of waves are excited, either the absorption or the reflectivity in the plasma can be enhanced.

In Chapter 3 we saw that three different wave modes of propagation are possible in an ICF plasma:

- electromagnetic waves
- electron waves, which is also denoted plasma (sometimes plasmon) or Langmuir waves

e.m. $\longrightarrow$ ion wave + electron wave (decay instability)
e.m. $\longrightarrow$ electron + electron wave (two-plasmon decay instability)
e.m. $\longrightarrow$ e.m. + ion wave (stimulated Brillouin scattering)
e.m. $\longrightarrow$ e.m. + electron wave (stimulated Raman scattering)

where e.m. stands for electromagnetic wave.

Because energy and momentum have to be conserved, a wave-number and frequency matching criterion (Manley-Rowe relations) of the kind

$$\omega_0 \approx \omega_1 + \omega_2 \qquad \text{and} \qquad \mathbf{k}_0 \approx \mathbf{k}_1 + \mathbf{k}_2$$

has to be fulfilled, where ω_0 and k_0 indicate the incident laser light and the indices 1 and 2 stand for the different types of decay modes.

It follows that these instabilities only occur if a certain relation between the incident laser light and the plasma frequency are fulfilled. For the above four cases this implies the following selection rules:

Instability type	Wavelength condition
Decay	$\omega_L \approx \omega_p$
Two-electron wave	$\omega_L \approx 2\omega_p$
Stimulated Brillouin	$\omega_p < \omega_L < 2\omega_p$
Stimulated Raman	$\omega_L > 2\omega_p$

Since the plasma frequency is determined by the plasma density, these conditions correspond directly to a location in the plasma (see Fig. 4.1). The decay instability and stimulated Brillouin scattering (SBS) occur at the critical density $n_e \sim n_c$, whereas the two-plasmon decay and stimulated Raman scattering occur near $n_e \sim n_c/4$.

However, even if these conditions on the wavelength are fulfilled, these instabilities are not always present in a plasma. The intensity of the incoming laser wave has to exceed a certain threshold in order for the parametric instability to occur, because all natural oscillation modes are damped. If the laser intensity is higher than this limit, the amplitude of the parametric decay mode increases exponentially with a characteristic growth rate, thereby absorbing laser energy. The frequency of the amplified oscillation is determined by the laser frequency rather than the natural frequency of these modes — this effect is often referred to as frequency locking.

The minimum laser energy necessary to invoke an instability can be calculated the following way:

$$\frac{}{dt^2} + 2\Gamma_1\frac{}{dt} + (\omega_1 + \Gamma_1)x(t) = \lambda z(t)y(t),$$

$$\frac{d^2y}{dt^2} + 2\Gamma_2\frac{dy}{dt} + (\omega_2^2 + \Gamma_1^2)y(t) = \lambda z(t)x(t),$$

where $x(t)$ and $y(t)$ are the oscillator amplitudes and $z(t) = 2z_0\cos\omega_0 t$ represents the laser amplitude. The coefficients Γ_1 and Γ_2 describe the degree of damping.

For electrostatic waves in a plasma these equations have to be applied in the following way. The laser beam is described as a uniform oscillating electric field and the coupled hydrodynamic equations are:

$$\frac{\partial n_j}{\partial t} + \mathbf{v}_j\frac{\partial n_j}{\partial \mathbf{r}} + n_j\frac{\partial}{\partial \mathbf{r}}\mathbf{v}_j = 0,$$

$$n_j\left(\frac{\partial \mathbf{v}_j}{\partial t} + \mathbf{v}_j\frac{\partial v_j}{\partial \mathbf{r}}\right) + \frac{1}{m_j}\frac{\partial p_j}{\partial \mathbf{r}} = \frac{e_j n_j}{m_j}\mathbf{E} - \nu_j n_j \mathbf{v}_j$$

$$\frac{\partial}{\partial \mathbf{r}}\mathbf{E} = \frac{1}{\epsilon_0}\sum_j e_j n_j. \tag{4.27}$$

There exist such a set of equations each for the electrons and ions indicated by the subscript j. To simplify things here, these equations are linearized to determine the threshold and growth rate. The nonlinear analysis is much more complex and usually requires numerical simulation.

Spatial linearization and averaging over the high-frequency motion of the electrons simplifies equations 4.27 to

$$\frac{\partial^2 n_e}{\partial t^2} + \nu_e\frac{\partial n_e}{\partial t} + \omega_{pe}^2(k)n_e = \frac{ie}{m_e}\mathbf{k}\cdot\mathbf{E}_0 n_i$$

$$\frac{\partial^2 n_i}{\partial t^2} + \nu_i\frac{\partial n_i}{\partial t} + \omega_{ia}^2(k)n_i = \frac{ie}{m_i}\mathbf{k}\cdot\mathbf{E}_0 n_e,$$

where $\omega_{ia} = (k_B T/m_i)^{1/2}k$ is the ion acoustic wave frequency. This set of equations is equivalent to that of coupled oscillators shown above. The dispersion relation corresponding to this set of equations can be obtained by Fourier transformation for the frequency matching condition $\omega_0 = \omega_1 + \omega_2$ as

$$\omega^2 + i\nu_i\omega - \omega_{ia}^2 = \frac{\omega_{pe}^2\omega_{pi}^2 k^2 e E_0}{4m\omega_0^2}\left(\frac{1}{\delta\omega^2 - \omega_R + i\nu_e\delta\omega} + \frac{1}{\omega_a^2 - \omega_R + i\nu_e\omega_a}\right),$$

$$\tag{4.28}$$

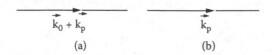

(a)   (b)

Figure 4.8. Raman scattering: a) forward and b) backward scattering.

where $\delta\omega = \omega - \omega_0$, $\omega_a = \omega + \omega_0$ and the electron plasma frequency $\omega_R = \omega_{pe} + \gamma_e k_B T_e k^2/m_e$. This dispersion relations is obviously very complex, so we look at the two distinct processes of stimulated Raman and Brillouin scattering separately.

Raman Scattering

Stimulated Raman scattering (SRS) is the resonant decay of the incident light into a scattered light wave and an electron plasma wave, with frequency and wave number matching conditions

$$\omega_0 = \omega_1 + \omega_2$$
$$k_0 = k_1 + k_2,$$

where ω_0 is the frequency of the incident light, ω_1 that of the scattered light and ω_2 is approximately ω_p, the electron plasma frequency. Figure 4.8 shows diagrams of the two cases of forward and backward SRS.

Why does an instability arise from this decay process? The answer is that a feedback loop develops: if small density fluctuations δn exist in the plasma, the electrons which oscillate in the presence of the electric field E_L of the laser light oscillate and produce a transverse current. The combination of the scattered light wave with the incident field increases the density fluctuations through the ponderomotive force ($\sim E_L E_S$). In this instability the plasma wave and the scattered light wave grow while weakening the incident lightwave: light can be scattered forward as well as backward.

As the frequency of the scatter light is higher than that of the electron plasma wave, $\omega_s > \omega_{ep}$, the instabilities occur at densities smaller than $1/4n_c$. The maximum growth occurs for backscattered light, with a growth rate $\gamma_{\text{Raman}}^{\text{back}}$ given by

$$\gamma_{\text{Raman}}^{\text{back}} = \frac{1}{4}k v_{osc}\left(\frac{\omega_p}{\omega_b}\right)^{1/2}, \tag{4.29}$$

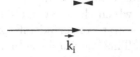

Figure 4.9. Schematic picture of Brillouin scattering.

where k is the wave number of the electron plasma wave and v_{osc} the oscillatory velocity of an electron in the laser field. The forward scattered light has the smaller growth rate of

$$\gamma_{\text{Raman}}^{\text{for}} = \frac{\omega_p^2 v_{osc}}{2\sqrt{2}\omega_0 c} \tag{4.30}$$

and is therefore less dangerous for the ICF process. If the density is close to $1/4n_c$ the wave number k_{ep} is similar to that of the incident wave, but if the density is much smaller than $1/4n_c$ then $k_{ep} \sim 2k_1$. If the plasma density becomes very low, SRS is suppressed by Landau damping of the plasma wave, which is the case when $k_{ep}\lambda_D > 1/3$. Stabilization may also occur collisional damping in a high-Z plasma.

Brillouin Scattering

The feed back mechanism of stimulated Brillouin scattering is similar to that of SRS. The difference now is that the density fluctuation is associated with a low-frequency ion acoustic wave (see Fig. 4.9). This time the wave number matching condition reads

$$\omega_0 = \omega_s + \omega_{pi}$$

and the maximum growth rate is given by

$$\gamma_{\text{Brill}}^{\text{back}} = \frac{\sqrt{3}}{2}\left(\frac{k_0^2 v_{osc}^2}{2}\frac{\omega_{pi}^2}{\omega_0}\right)^{1/3}. \tag{4.31}$$

This can lead to high reflectivity. Nearly all laser plasma instability growth rates increase as a function of $v_{osc} \sim (I\lambda^2)^{1/2}$, and for this reason short wavelength lasers (0.35 μm, 0.248 μm) are strongly preferred as drivers for ICF.

laser directly irradiates the target. For indirect-drive targets there is an additional step to consider in which the laser irradiates the hohlraum wall instead, and the laser light is first converted into x-ray radiation, which then drives the compression. This inevitably entails an additional loss of energy — namely the amount that is absorbed in the wall rather than re-emitted from the hohlraum walls.

To determine how much energy is absorbed or re-emitted into the hohlraum itself, we use a simplified approach, assuming a planar hohlraum wall surface, thus allowing a one-dimensional description. In addition, it will be assumed that radiation and matter are in local thermal equilibrium. From the conservations laws of mass, momentum and energy (Eq. 3.4), one obtains the following set of fluid equations in Lagrangian form:

$$\frac{\partial}{\partial t}\frac{1}{\rho} = \frac{\partial v}{\partial m},$$

$$\frac{\partial v}{\partial t} = \frac{\partial P}{\partial m},$$

$$\frac{\partial \epsilon}{\partial t} + P\frac{\partial}{\partial t}\frac{1}{\rho} = -\frac{\partial F}{\partial m}. \tag{4.32}$$

Here, m is the mass of the matter between the fluid particle and the surface wall per unit area, v is the fluid velocity in the rest frame, P the pressure, and F the energy flux. In the energy equation we use the fact that heating of a fluid particle TdS is related to the energy transported out of a fluid element by $TdS = d\epsilon + Pd(1/\rho)$. The dominant heat-transport mechanism in this situation is radiation transport. It can be assumed that the opacity is high enough that the diffusion approximation holds, so that the energy flux can be described by

$$F = -\frac{4\sigma_{SB}}{3K_R}\frac{\partial T^4}{\partial m}, \tag{4.33}$$

where σ_{SB} is the Stefan–Boltzmann constant and K_R is the Rosseland mean opacity. If the temperature suddenly induced by the laser irradiation at $t = 0$ is T_0, and we can make some estimates of the diffusion process.

Early in the interaction process, the thermal wave proceeds much faster than the hydrodynamic wave. As the laser intensity is so high at early times, the material density stays nearly constant i.e., $(\partial/\partial t(1/\rho) = 0)$. It follows from Eqs. 4.32 and 4.33 that

$$\frac{\partial \epsilon}{\partial t} = -\frac{\partial}{\partial m}\frac{4\sigma_{SB}}{3K_R}\frac{\partial T^4}{\partial m} \tag{4.34}$$

The subscript 0 denotes the values at $t = 0$. From Section 3.7 we know that hydrodynamic disturbances move with the sound speed; in our notation, this is equivalent to

$$\left(\frac{m}{t}\right)_h \approx \sqrt{P_0 \rho_0}. \tag{4.36}$$

Equations 4.35 and 4.36 can be used to estimate when and where the hydrodynamical speed and the thermal speed become comparable. The time $t_{h=th}$, when these two speeds are approximately the same, is given by

$$t_{h=th} = \left(\frac{4\sigma_{SB} T_0^4}{3 K_{R\epsilon 0}} \frac{1}{P_0 \rho_0}\right).$$

At this time the hohlraum has been heated to a depth

$$m_{h=th} = t_{h=th} \sqrt{P_0 \rho_0}.$$

At times $t \gg t_{h=th}$ the situation becomes more complex. At this point the hydrodynamics begins to influence the heat front and the assumption above is no longer valid. Numerical simulations are necessary to solve the complete equation system.

In the case of a hohlraum target typical for the planned NIF experiments (see, for example Lindl, 1995) the energy absorbed in the gold hohlraum wall after a time τ can be approximated by an empirical relation:

$$E_{wall}(\text{MJ}) - 5.2 \times 10^{-3} K_0^{-0.39} [3.44 P + 1]^{-0.39} \times T_0^{3.3} \tau^{0.62 + 3.3 P} A_{wall},$$

where T_0 is the temperature at 1 ns and A_{wall} the wall area of the hohlraum. For a constant hohlraum temperature (i.e., $T_0 = T_{hohl}$ and $P = 0$), it follows

$$E_{wall}(\text{MJ}) = 5.2 \times 10^{-3} K_0^{-0.39} T_{hohl}^{3.3} \tau^{0.62} A_{wall}. \tag{4.37}$$

The energy loss per time is then given by the time derivative of Eq. 4.37, that is

$$\frac{dE_{wall}}{dt}(\text{MJ/ns}) = \frac{5.2 \times 10^{-3}}{0.62 + 3.3 P} K_0^{-0.39} T_{hohl}^{3.3} A_{wall} \tau^{-0.38 + 3.3 P}.$$

For a constant loss rate, the condition $t_0 = \tau^{-0.38 + 3.3 P}$ has be to fulfilled, which means $1 = -0.38 + 3.3 P$ and therefore $P = 0.115$. For the absorbed flux $S_{abs} = A_{wall}^{-1} dE_{wall}/dt$ it follows

$$S_{abs}(\text{MJ/ns/cm}^2) = 4.5 \times 10^{12} T_{hohl}^{3.3} K_0^{-0.39}. \tag{4.38}$$

absorbed power.

The black-body radiation flux *emitted* from the hohlraum wall is about

$$S_{rad}(\text{MJ/ns/cm}^2) \sim 10^{13}T^4.$$

Because $T = T_{hohl}\tau^{0.015}$, it follows for the $T_{hohl}^{3.3}$ dependence in Eq. 4.38

$$T^{3.3} = T_0^{3.3}\tau^{0.38} \sim S_{rad}^{0.825}.$$

Expressing T_{hohl} in terms of S_{rad} and substituting it into Eq. 4.38, one obtains

$$S_{rad}[10^{15}\text{MJ/ns/cm}^2] = 7.0K_0^{0.47}S_{abs}^{1.21}(10^{15}\text{MJ/ns/cm}^2)\tau(\text{ns})\tau(\text{ns}).$$

For typical conditions in ICF hohlraums ($\tau \sim 1$ ns and fluxes of 10^{14} – 10^{15} W/cm^2) the percentage of the flux re-emitted from the hohlraum wall to the flux that absorbed by the wall is relatively high. Losses in the hohlraum wall are therefore tolerable.

As pointed out before, the above treatment assumes a constant laser pulse with a constant x-ray conversion efficiency. In the ICF scenario both the laser pulse and the x-ray conversion efficiency are time-dependent. In this case an analytical treatment is no longer possible. For a summary of the numerical and experimental results concerning the conversion of laser light into x-ray at hohlraum walls we refer to Lindl (1995).

Despite some losses through this x-ray conversion process, the indirect-drive scheme is nevertheless currently the preferred approach to ICF because the requirements for radiation uniformity are easier to achieve in indirect drive than in direct drive.

Before we continue with the description of the energy transport, we consider what happens to the hohlraum when it is heated from the inside: does it completely evaporate? Obviously when the laser hits the inside of the hohlraum, mass is ablated from the hohlraum wall, the amount of which can be approximated by (see Lindl, 1995)

$$m_{wall}^{abl} = 1.2 \times 10^{-3}T^{1.86}K_0^{-0.46} \times (3.44P + 1))^{-0.46}\tau^{0.54+1.86P}.$$

Assuming a constant loss rate $P=0.115$ as before, then the depth to which the laser penetrates is given by

$$x_{wall}(\mu\text{m}) = 10^4 m/\rho = 0.53T_0^{1.86}\tau^{0.75} \tag{4.39}$$

for a gold wall, which is a typical material for such a hohlraum. From Eq. 4.39 it follows that for times of around 1 ns and temperatures relevant for

discuss its influence on the overall power balance calculation.

4.6 Energy Transport

Here we return to the point where we left off in Section 4.4 in which we saw how the energy is absorbed by various processes. The question now is how this absorbed energy gets transported from the critical surface toward the solid-density fuel. There are two mechanisms to transport the energy — radiation and thermal conduction. We will first consider the latter.

Electron Thermal Conduction

The thermal conduction process is dominated by the much lighter and faster electrons, whereas the slow heavy ions can be largely. To first order one can treat the process as electrons diffusing through a background of fixed ions.

In this classical description of electron thermal conduction the heat flux $\mathbf{q_{th}}$ is given by

$$\mathbf{q}_{th} = -\kappa \nabla T, \qquad (4.40)$$

where κ is the thermal conductivity, which can be estimated by a simple model. For a flux of particles in direction x with an average energy $\epsilon(x)$, the heat flux in the direction x can be approximated as

$$q_{th}(x) \sim -\frac{1}{3}nvl\frac{\partial \epsilon}{\partial x} = -\frac{1}{3}nvl\frac{\partial \epsilon}{\partial T}\frac{\partial T}{\partial x} = -\frac{1}{3}nvlc_v\frac{\partial T}{\partial x}, \qquad (4.41)$$

where l is the mean free path and c_v the heat capacity per particle per volume. The factor $1/3$ reflects the averaging over the three space dimensions. Assuming that the velocity is the thermal velocity, the thermal conductivity is given by (Spitzer and Harm, 1953)

$$\kappa = \frac{1}{3}\lambda_e v_e n_e k_B = \frac{5n_e k_B^2 T_e}{m_e \nu_{ei}} = 20\left(\frac{2}{\pi}\right)^{3/2}\frac{(k_B T_e)^{5/2}k_B}{m_e^{1/2}e^2 Z.\ln\Lambda} \qquad (4.42)$$

From kinetic theory if follows that heat flow is only possible if the Maxwellian velocity distribution is in some way distorted. This means that there has to be a fraction of hot electrons which carry the thermal energy. The flow of the hot electrons creates an electric field $\mathbf{E}$, which in turn induces a return drift of cold electrons (see Fig. 4.10). Because the hot electrons from

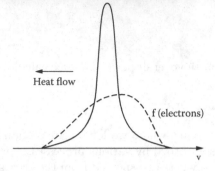

Figure 4.10. Skewed electron velocity distribution necessary for electron thermal conduction.

the ablation front have a high energy and the cold electrons of the return current a low energy, the target is heated. The return current means that a better approximation to the net heat flow contains an additional term and reads

$$\mathbf{q}_{th} = -\kappa \nabla T - \beta_P \mathbf{E}, \qquad (4.43)$$

where β_P is the so-called Peltier coefficient. The electric current $\mathbf{j}$ is then given by

$$\mathbf{j} = \sigma_E \mathbf{E} - \alpha \nabla T. \qquad (4.44)$$

Demanding that there should be no net current(i.e., $\mathbf{j} = 0$), combining Eqs. 4.43 and 4.44 it follows

$$\mathbf{q}_{th} = -\kappa \left(1 - \frac{\alpha \beta_P}{\sigma_e \kappa} \right) \nabla T. \qquad (4.45)$$

with an effective $\kappa_{\text{eff}} = (1 - (\alpha \beta_P)/(\sigma_e \kappa))\kappa = \delta_e \kappa$, the general form of the electron thermal conductivity is given by

$$\kappa = \delta_e 20 \left(\frac{2}{\pi} \right)^{3/2} \frac{(k_B T_e)^{5/2} k_B}{m_e^{1/2} e^2 Z \ln \Lambda}. \qquad (4.46)$$

Comparing Eqs. 4.42 and 4.46 it can be seen that the return current effectively reduces the heat transport. This reduction can be up to a factor of 2.

If the temperature gradient is very steep the classical diffusion approach is no longer valid. It has been shown by Gray and Kilkenny (1980) that the diffusion approximation is valid as long as

$$\lambda_e \frac{|\nabla T|}{T} > 0.01. \qquad (4.47)$$

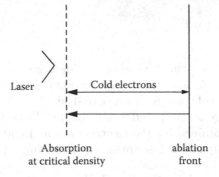

Figure 4.11. Schematic picture of electron thermal conduction.

If the electron mean-free-path is larger than the temperature gradient, Eq. 4.45 no longer holds. In computer simulations one usually introduces a so-called thermal flux limiter to interpolate between Eq. 4.45 and the limit of a free streaming flux. In many simulation codes an expression such as:

$$q = \left[\frac{1}{\kappa_e |\nabla T_e|} + \frac{2}{mv^2 n_e v} \right]^{-1}$$

is used to handle this problem in practice.

Thermal Conduction Inhibition

The actual measured thermal flux in coronae is nearly an order of magnitude smaller than the predicted by the above model, especially for steep gradients (i.e., $\lambda_e \nabla T/T < 0.01$). In practice, this transport inhibition is usually treated by introducing an inhibition parameter f_{inhib}. Equation 4.46 then reads

$$\kappa = f_{inhib} \delta_e 20 \left(\frac{2}{\pi} \right)^{3/2} \frac{(k_B T_e)^{5/2} k_B}{m_e^{1/2} e^2 Z \ln \Lambda}. \tag{4.48}$$

Simulations show that an inhibit factor of 0.03 is needed to reproduce experimental results (Rosen, 1984).

However, this is not a very satisfying situation: one should at least know which processes are responsible for this conduction inhibition, usually connected to the strong density and temperature gradients. The reasons for the conduction inhibition is the strong dependence of the collisional mean free path on the particle kinetic energy. This limits the validity of the above described standard local transport relations to very long inhomogenity scale length l_{inh}.

lasers. In this case the hydrodynamic description no longer suffices, and the Fokker-Planck description (see Section 3.2) is required.

Using the solution to a linearized kinetic equation and assuming a large ion charge $Z \gg 1$ and small amplitude perturbations, Bychenkov et al. (1995) obtained for the electric current $\mathbf{j}$ and the electron heat flux $\mathbf{q}_e$ for a hydrodynamic description the following expressions:

$$\mathbf{j} = \sigma \mathbf{E}^* + \alpha_{th} i \mathbf{k} \delta T_e + \beta_j e n \mathbf{u}_i$$
$$\mathbf{q}_e = - \alpha_{th} T_0 \mathbf{E}^* - \chi_c i \mathbf{k} \delta T_e - \beta_q n_0 t_0 \mathbf{u}_i \qquad (4.49)$$

where $\mathbf{E}^* = -i\mathbf{k}\Phi + (i\mathbf{k}/en_0)(\delta n T_0 + \delta T n_0)$ and the electric conductivity σ, the thermoelectric coefficient α_{th}, the temperature conductivity χ_c and the ion density flux coeffcient $\beta_{j,q}$ given by

$$\alpha = - (en_0/m_e k^2 v_{Te}^2)[-p + (J_T^N + J_T^T)/D_{NT}^{NT}]$$
$$\chi = (n_0/k^2)[-5p/2 + (2J_T^N + J_T^T + J_N^N)/D_{NT}^{NT}]$$
$$\sigma = (e^2 n_0/m_e k^2 v_{Te}^2)(-p + J_T^T/D_{NT}^{NT})$$
$$\beta_q = (D_{NT}^{RT} + D_{NT}^{RN})/D_{NT}^{NT}]$$
$$\beta_j = 1 - D_{NT}^{RT})/D_{NT}^{NT}.$$

The transport relations 4.49 exhibit the so-called Onsager symmetry: The coefficent α_{th} is the same in both expressions for $\mathbf{j}$ and $\mathbf{q}_e$.

Measurements of spatially and temporally resolved temperature and density profiles by Gregori et al. (1942) find that the profiles disagree with flux-limited models, but are consistent with nonlocal transport modeling.

In cases in which the above assumption cannot be made, a Fokker-Planck treatment is required.

Preheat

As mentioned in Chapter 3, the energy transfer from electron plasma waves, nonlinear wave phenomena, and resonance absorption can create high-energy electrons, so-called hot electrons. These hot electrons can move to the center of the target in front of the ablation front, thus preheating the fuel and making it less compressible. In hohlraum targets Raman instabilities produce hot electrons when the plasma waves are damped in the hohlraum.

Experiments with planar targets show that the hot electron temperature depends on the laser power the following way for $I_L \lambda_L^2 < 10^{15}$ W cm$^{-2} \mu$m^2:

$$T_h[\text{keV}] = 10 \left(\frac{I_L \lambda_L^2}{10^{15} \text{ W cm}^{-2} \mu\text{m}^2} \right)^{2/3}. \qquad (4.50)$$

The main problem with the production of hot electrons is the reduced compressibility of the preheated fuel. Modern targets and beam shapes are designed in such a way that the preheat ϵ_{pre} should not exceed the specific Fermi energy ϵ_F of the fuel during the implosion,

$$\epsilon_{pre} < \epsilon_F. \qquad (4.51)$$

It is essential that this condition is fulfilled, because even in the case $\epsilon_{pre} = \epsilon_F$, the pressure needed to obtain a given density doubles in comparison to a compression without preheat.

In such a plasma the electron velocity distribution cannot be described by a single Maxwellian distribution. A superposition of two Maxwellian distributions with two temperatures — the mean temperature T_c of the cold electrons and the mean temperature T_h of the hot electrons — is generally a much better description.

Electrons at the lower end of the electron velocity distribution do not reach the fuel, but how many hot electrons reach the fuel? This depends on thickness x of the ablator and the mean range x_0 of the hot electrons,

$$\epsilon_{pre}(x) = \epsilon_{x=x_0} G(x/x_0), \qquad (4.52)$$

where $G(x/x_0)$ is an attenuation factor. The mean range $x_0[\text{g/cm}^2]$ of hot electrons of a temperature $T_h[\text{keV}]$ is given by

$$x_0 = \frac{3 \times 10^{-6}(A/Z)T_{hot}^2}{Z^{1/2}}. \qquad (4.53)$$

If one assumes that the hot electron velocities are Maxwellian, the preheat $\epsilon_{pre}(x)[\text{J/g}]$ caused in an area $A[\text{cm}^2]$, depends on the thickness x of the ablator and the energy E_{hot} of the hot electrons according to:

$$\epsilon_{pre}(x) = \frac{E_{hot}}{xA} \left(\frac{x}{x_0} \right)^{0.9} \exp\left[-1.65 \left(\frac{x}{x_0} \right)^{0.4} \right]. \qquad (4.54)$$

The tolerable fraction f_{hot} of the laser energy into hot electrons depends on the Fermi energy ϵ_F, the attenuation factor $G(x/x_0)$, and on the fuel

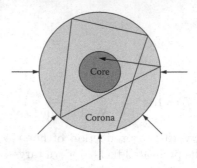

Figure 4.12. Scattering of hot electrons in target.

mass m_{fuel} and the laser energy E_L:

$$f_{hot} = \frac{\epsilon_F m_{fuel}}{E_L G(x/x_0)}. \tag{4.55}$$

An effect that has been neglected in this estimate is that the hot electrons see a relatively transparent plasma except for the very dense core of the target. Therefore, if the hot electrons are isotropic, in both direct- and indirect-drive a considerable amount of these hot electrons do not hit directly the core but reach it after many scattering events as illustrated in Fig. 4.12. If we include these effects in Eq. 4.55 by a so-called geometric dilution factor $D(r_{source}/r_{cap})$, the tolerable fraction is then:

$$f_{hot} = \frac{\epsilon_F m_{fuel}}{E_L G(x/x_0) D(r_{source}/r_{cap})}. \tag{4.56}$$

However, there is some difference as how this effect manifests itself in direct- and indirect-drive. In direct-drive targets this scattering occurs off the space charge potential in the outer regions of the corona (see Fig. 4.12). When the hot electrons hit the core, they penetrate in front of the shock wave ablation region.

In indirect-drive targets, only those electrons that have velocities in direction of the capsule angle will hit the capsule directly. Because most hot electrons will be produced on average at the hohlraum case radius, the proportion of the hot electrons hitting directly the capsule will be approximately $A_{case}/A_{capsule}$, where A_{case} is the case area (i.e., the inside area of the hohlraum, and $A_{capsule}$ the capsule area). Hot electrons that miss the capsule can be reflected off the case. Experiments show that the likelihood of the hot electrons being reflected by the high-Z case is about

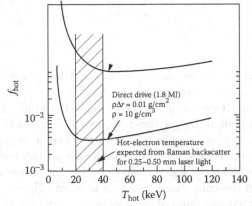

Figure 4.13. Comparison of preheat effects in direct and indirect drive targets. The tolerable fraction of hot electrons f_{hot} is shown as a function of the hot electron temperature. In these calculations it was assumed that the fuel preheat is equal to the Fermi energy. Reprinted with permission from Lindl (1995) ©1995, American Institute of Physics.

50% with about 70% of their energy (Lindl, 1995). These reflected hot electrons might then hit the capsule.

Overall, direct-drive targets are more sensitive to preheat than indirect-drive targets.

- In direct-drive targets the beams are absorbed very close to the ablator, so the geometry does decrease the number of hot electrons that hit the core too a lesser extend.
- In direct-drive targets there is less ablator material to shield the fuel from preheat.

Figure 4.13 shows a comparison of simulations for the direct- and indirect-drive scheme studying the effect of hot electrons for capsules typical in national ingition facility (NIF) experiments. These simulations assumed a fuel preheat of the order of the Fermi energy. The fraction of hot electrons f_{hot} that would be tolerable for the target to still reach ignition is much less in the direct-drive scenario than in indirect-drive case. As Fig. 4.13 shows, the tolerable fraction of hot electrons lies direct-drive targets lies below 1%, whereas for indirect-drive 5% of hot electrons is still tolerable.

However, this might be too pessimistic for direct-drive targets. For direct-drive capsules an adiabat at 3–4 times the Fermi energy is probably more suitable as this would reduce hydrodynamic instabilities. In this case the tolerable fraction of hot electrons f_{hot} could equally be a factor of 3–4 higher, making both schemes comparable again.

Olson *et al.* (2003) showed that in indirect drive ablator preheat can

When the laser interacts with the plasma, x-ray radiation is produced (Gauthier, 1989; Kauffman, 1991; More *et al.*, 1988). The x-rays are produced by various electron transitions, namely, bound-bound, bound-free and free-free transitions. The probability of these processes is proportional to the density squared (i.e., n^2), so most x-rays are produced in regions of high electron density. Therefore a short laser wavelength of the incident laser light is advantageous, because shorter wavelength light reaches to higher electron densities and the conversion to x-rays is more efficient.

Naturally the actual spectrum of the produced x-ray radiation depends on the ablator material, the plasma density, and temperature, and is a superposition of the lines and continuous energies emitted because a result of transitions. In high-Z materials the laser energy is much more efficiently converted into x-rays than in low-Z materials, which is why high-Z materials such as gold are used for the hohlraum ablator surface. Experiments have shown that in gold targets 80% of the energy of a 1-ns long 263 nm laser pulse was converted into x-rays in the energy range of 0.1–1 keV.

To describe the radiation transport, we first will define some quantities:

- The spectral radiation intensity $I_\nu [\text{erg/cm}^2/\text{s}]$ is usually defined as the radiation energy transported per frequency between ν and $\nu + d\nu$ crossing a unit area per time in the direction Ω, within the solid angle $d\Omega$,

$$I_\nu(\mathbf{r}, \boldsymbol{\Omega}, t) = h\nu c f(\mathbf{r}, \boldsymbol{\Omega}, t) d\nu d\boldsymbol{\Omega},$$

   where h is the Planck constant, c the speed of light and f the photon distribution function.
- The emmisivity $j_\nu [\text{erg/cm}]^3$ describes the process of spontaneous emission and is defined as

$$j_\nu d\nu d\boldsymbol{\Omega} = [\text{energy/(volume} \times time)] \text{ of spontaneous emission.}$$

   This depends on the type of atom, the ionization degree, and the temperature of the plasma, but is independent of the existence of radiation.
- By contrast, for induced emission, which is given by

$$j_\nu \frac{c^2 I_\nu}{2h\nu^3} d\nu d\boldsymbol{\Omega} = [\text{energy/(volume} \times time)] \text{ of induced emission.}$$

   The existence of radiation is mandatory.

where κ_ν is the opacity given by $1/l_\nu \quad \sum_j n_j \sigma_{\nu j}$ with l_ν being the mean free path and $\sigma_{\nu j}$ the appropriate absorption and scattering cross sections.

Imposing a conservation of the energy density in a given frequency ν and direction ω in an arbitrary volume V and combining all above processes, the radiation transport equation can be written as

$$\frac{1}{c}\left(\frac{\partial I_\nu}{\partial t} + c\boldsymbol{\Omega} \cdot \nabla I_\nu\right) = j_\nu\left(1 + \frac{c^2 I_\nu}{2h\nu^3}\right) - \kappa_\nu I_\nu. \tag{4.57}$$

In thermal equilibrium the ratio of the spontaneous emission to the absorption is given by (see Section 2.1)

$$\frac{j_\nu}{\kappa_\nu} = \frac{2h\nu^3}{c^2} \exp\left(-\frac{h\nu}{k_B T}\right). \tag{4.58}$$

Using the following definitions of $I_{\nu p}$ and κ_ν',

$$I_{\nu p} = \frac{j_\nu}{\kappa_\nu} \frac{1}{\left[1 - \exp\left(-\frac{h\nu}{k_B T}\right)\right]},$$

$$\kappa_\nu' = \kappa_\nu\left[1 - \exp\left(-\frac{h\nu}{k_B T}\right)\right],$$

the transport equation can be rewritten as

$$\frac{1}{c}\frac{\partial I_\nu}{\partial t} + \boldsymbol{\Omega} \cdot \nabla I_\nu = \kappa_\nu'(I_{\nu p} - I_\nu). \tag{4.59}$$

To analyze this transport equation, one takes the angular moments. The continuity equation is obtained as

$$\frac{\partial u_\nu}{\partial t} + \nabla \mathbf{S}_\nu + c\kappa_\nu' u_\nu = c\kappa_\nu' u_{\nu p} \tag{4.60}$$

by integrating over solid angles. Here $u_{\nu p}$ is the equilibrium radiation energy density

$$u_{\nu p} = \frac{8\pi h\nu^3}{c^3} \frac{1}{e^{h\nu/kT} - 1} \tag{4.61}$$

and $\mathbf{S}_\nu$ the radiation flux. Higher order moments are obtained by multiplying with $\boldsymbol{\Omega}$ and angular integration. The next higher moment is then given by

$$\frac{1}{c}\frac{\partial \mathbf{S}_\nu}{\partial t} + \nabla \boldsymbol{\Gamma}_\nu + \kappa_\nu' \mathbf{S}_\nu - 0, \tag{4.62}$$

However, if I_ν is nearly isotropic, Eq. 4.60 suffices and higher moments can be neglected.

In this case the radiation energy density can be related to the energy flux in the diffusion approximation via:

$$\mathbf{S} = -\frac{c}{\kappa'_\nu} \nabla u_\nu \tag{4.64}$$

and the diffusion equation of the spectral radiant energy density reads

$$\frac{\partial u_\nu}{\partial t} - \nabla \frac{c}{\kappa'_\nu} \nabla u_\nu + c\kappa'_\nu u_\nu = c\kappa'_\nu u_{\nu p} \tag{4.65}$$

or if we use the expression

$$J_\nu = c\kappa_\nu u_{\nu p} \tag{4.66}$$

for the radiant energy density J_ν, it can be expressed as

$$\frac{\partial u_\nu}{\partial t} - \nabla \frac{c}{\kappa_\nu} \nabla u_\nu + c\kappa'_\nu u_\nu = J_\nu. \tag{4.67}$$

This equation can in principle be solved by any standard method for diffusion problems. However, the difference to conventional heat transport is that all the coefficients — emission, absorption, and diffusion coefficient — are not constant, but strong functions of the plasma temperature.

A second difference to other diffusion processes is that the photons "diffuse" through a number of absorption and reemission processes. If the radiation diffuses into a colder plasma area, the area is heated and the properties of the diffusion process are changed. To describe the diffusion process self-consistently the radiation diffusion equation has to be solved in tandem with the plasma hydrodynamics, leading to a highly complex system of equations which must in general be solved numerically.

However, a rough estimate of the relative importance of radiative and thermal transport can be obtained the following way:
The radiation heat flux is given by

$$\mathbf{q}_{rad} = -\frac{\lambda_R c}{3} U_p \tag{4.68}$$

where λ_R is the radiation mean-free-path. The radiation energy is

$$U_p = \frac{4\sigma_{SB} T^4}{c}. \tag{4.69}$$

$$\mathbf{q}_{rad} = -\kappa_R \nabla T. \qquad (4.71)$$

For a hydrogen-like plasma with a charge Z the mean free path can be approximated in the Rosseland case by

$$\lambda_R(\text{H-like}) \sim 8.7 \times 10^6 \frac{T^2}{Z^2 n_i} \qquad [\text{cm}]. \qquad (4.72)$$

Comparing the radiation thermal conductivity with the electron thermal conductivity κ_e (Spitzer–Härm conductivity), it follows

$$\frac{\kappa_R}{\kappa_e} \sim 5 \times 10^{18} \left(\frac{k_B T}{\text{eV}}\right)^{5/2} \left(\frac{\text{cm}^{-3}}{n_e}\right). \qquad (4.73)$$

This means for a solid state density plasma ($n_e \sim 10^{23}$ cm^{-3}) the threshold where radiation transport becomes more important than electron heat transport lies at $k_B T > 100$ eV. Thus in cold dense plasmas, electron transport dominates, whereas at later stages when the plasma is already heated to some degree, radiation transport takes over.

Chapter 5

Hydrodynamic Compression and Burn

After the laser light has been absorbed, the energy transport and the ablation processes set off the next step in the inertial confinement fusion (ICF) process — the hydrodynamical compression phase. The parameter most critical for the entire success of ICF ignition is the implosion velocity achieved in this step.

In this chapter we will first compare a target in which the whole sphere is filled with fuel, with a target in which the fuel is concentrated in a thin shell. Determining the implosion velocities that can be achieved in either case, it can be demonstrated that it is very advantageous to use a shell (or foil) target configuration. We start by using a simple one-dimensional model with planar geometry and generalize this to spherical geometry.

In Chapter 1 it was explained why the hot-spot concept is the preferred

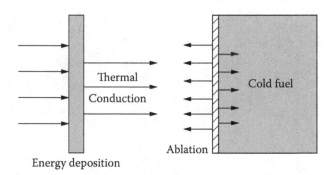

Figure 5.1. Different processes involved in target implosion.

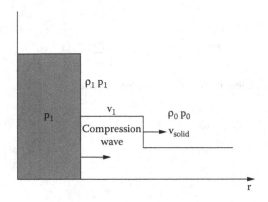

Figure 5.2. Compression of a homogenous medium for a "solid" target.

compression scheme for keeping the energy input for driving the hydrodynamic compression as low as possible. In this scenario, only a small central part of the compressed fuel is brought to the temperature necessary for ignition, whereas the rest of the fuel stays at a much lower temperatures (see Fig. 1.8). It is ignited by a thermonuclear burn wave propagating outward from the central spark.

5.1   Implosion of Solid Target

We start with the simple model of a solid target in a planar geometry illustrated by Fig. 5.2. Because of the applied laser field, a pressure p is generated that creates a shock wave propagating with constant velocity v_{solid}. If the coordinate system is chosen so that it moves with the shock wave, the three relevant conservation relations are:

Mass
$$\rho_0 v_{solid} = \rho_1 v_1,$$

Momentum
$$p_0 + \rho_0 v_{solid}^2 = p_1 + \rho_1 v_1^2,$$

Energy
$$\frac{\gamma}{\gamma - 1}\frac{p_0}{\rho_0} + \frac{v_{solid}^2}{2} = \frac{\gamma}{\gamma - 1}\frac{p_1}{\rho_1} + \frac{v_1^2}{2},$$

where the subscripts 0 and 1 denote quantities in front of and behind the shock wave, respectively, and γ describes the specific heat ratio of the materials. Assuming that there is no preheat, the material in front of the

$$\frac{\gamma}{\gamma-1}\frac{p_0}{\rho_0} + \frac{v_{solid}}{2} = \frac{\gamma}{\gamma-1}\frac{p_1}{\rho_1} + \frac{v_1}{2}.$$

This can be transformed into the convenient form:

$$\frac{\rho_1}{\rho_0} = \frac{\gamma+1}{\gamma-1}$$

$$v_1 = \frac{\gamma-1}{\gamma+1}v_{solid}$$

$$p_1 = \frac{2}{\gamma+1}\rho_0 v_{solid}^2.$$

From these equations the shock velocity can be expressed in terms of the applied pressure p_1:

$$v_{solid} = \left[\frac{(\gamma+1)}{2}\frac{p_1}{\rho_0}\right]^{1/2}. \tag{5.1}$$

For a massive planar target, the shock wave would propagate with this constant velocity until it reaches the center after a distance R. This velocity would obviously not suffice to obtain the neccessary compression.

5.2   Foil Target

In the next step we wish to see what implosion velocity can be obtained by using a target where the fuel sits in a layer at the inside of the capsule. In planar geometry this is equivalent to a foil of a certain thickness ΔR. If the foil thickness ΔR is small compared with the radius of the target R (i.e., $\Delta R \ll R$), the transversal time of the shock wave through the foil is negligible. Therefore one can treat the foil as if the entire material is heated instantly.

Assuming the foil remains rigid (an oversimplification — in reality the foil expands) its motion is given by

$$M_{foil}\frac{dv_{foil}}{dt} = -p_1 S, \tag{5.2}$$

where S is the surface and M_{foil} is the mass of the foil. This foil mass can be expressed as $M_{foil} = \rho_0 \Delta R S$, and substituting this into Eq. 5.2, one obtains

$$\frac{dv_{foil}}{dt} = \frac{p_1}{\rho_0 \Delta R}.$$

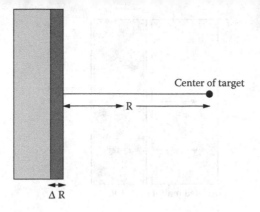

Figure 5.3. Acceleration of a foil target.

If the pressure p_1 remains constant, the implosion velocity is given by

$$v_{foil} = \frac{p_1 t}{\rho_0 \Delta R}.$$

Here, in contrast to the solid target case, the implosion velocity *increases* with time until the target center is reached. Therefore it follows from

$$R = \int v_{foil}\, dt = p_1 t^2/2\rho_0\Delta R$$

that

$$t = \sqrt{2R\rho_0\Delta R/p_1}$$

and the final implosion velocity of the foil is given by

$$v_{foil} = 2\left(\frac{p_1}{\rho_0}\frac{R}{\Delta R}\right)^{1/2}. \tag{5.3}$$

Comparing the implosion velocities of the solid v_{solid} and the foil target v_{foil}, we have

$$\frac{v_{foil}}{v_{solid}} = \left(\frac{8}{\gamma+1}\right)^{1/2}\left(\frac{R}{\Delta R}\right)^{1/2}. \tag{5.4}$$

This means that the velocity obtained in a foil target is a factor of $(R/\Delta R)^{1/2}$ bigger than in a solid target. According to Eq. 5.4 a large target with a very thin layer of fuel would be ideal — a so-called "high-aspect ratio target." We will see later that other processes such as Rayleigh–Taylor instabilities set upper limits on the total size as well as $R/\Delta R$.

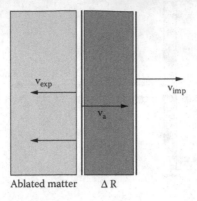

Ablated matter ΔR

Figure 5.4. Velocities in the rocket model: v_{exp} is the expansion velocity of the ablated material, v_{abl} is the velocity at which the ablation propagates and v_{imp} is the velocity of the implosion towards the center of the target.

5.3   Rocket Model and Ablation

The above foil model does not take into account the fact that part of the foil heats up and expands away from the target — a process known as ablation. In the following we will see that ablation is an essential feature in the compression of ICF capsules. A simple model that describes this effect is so-called "rocket-model." In a rocket the steady blow-off of the fuel accelerates the rocket. In the ICF process the conduction of heat into the ablation front builds up pressure which drives the ablation of matter from the outside of the capsule. This in turn leads to the acceleration of fuel in the opposite direction — that is, toward the center of the capsule — and drives the implosion of the target.

As the foil heats up and expands, the steady blow-off of ablated plasma at velocity v_{exp} leads to an acceleration of the target of mass m at velocity v_a. Taking the ablation surface of the target as our frame of reference as in Fig. 5.4, the three conservation laws read in this case

$$\frac{dm}{dt} = \rho v = const. \tag{5.5}$$

$$P_a = \rho v^2 + P \tag{5.6}$$

$$\left[\left(\frac{\gamma}{\gamma - 1} \right) \frac{dm}{dt} \left(\frac{P}{\rho} \right) \right] = - \left(\frac{1}{2} \right) \frac{dm}{dt} v^2 + q_{in} \tag{5.7}$$

For a monoatomic gas $\gamma = 5/3$, so that $\gamma/\gamma - 1 = 5/2$. The expansion speed v_{exp} of the plasma can be assumed to equal the sound speed c_s, which for a constant corona temperature is that of an isothermal gas (i.e.,

If we assume a constant rate of plasma production (i.e., dm/dt=constant), the mass is given as

$$m(t) = m_0 - \int_0^t \left(\frac{dm}{dt}\right) = m_0 - \dot{m}\, t, \tag{5.9}$$

where m_0 is the initial mass. Substituting Eq. 5.8 in Eq. 5.9 we find

$$(m_0 - \dot{m}\, t)\frac{dv}{dt} = 2\, \dot{m}\, v_{exp}.$$

Integration leads to the rocket equation

$$v(t) = v_{exp} \ln\left(\frac{m_0}{m}\right). \tag{5.10}$$

This means for the implosion velocity

$$v_{imp} = \frac{P}{\dot{m}} \ln \frac{m_0}{m}. \tag{5.11}$$

or

$$v_{imp} = v_{exp} \ln \frac{m_0}{m}. \tag{5.12}$$

What implosion velocities can we expect now for a given deposition of energy by a driver? As we saw earlier, v_{exp} is directly connected to the pressure and the mass ablation rate. The pressure generated by the ablation scales as a power of the incident flux:

$$P = P_0 I^{c_1} \tag{5.13}$$

as does the rate of ablation

$$\dot{m} = \dot{m}_0\, I^{c_2}. \tag{5.14}$$

For the implosion velocity it follows

$$v_{imp} = \frac{P}{\dot{m}} I^{c_1 - c_2} \ln \frac{m_0}{m}. \tag{5.15}$$

This is valid for direct- as well as indirect-drive. However the coefficients c_1 and c_2 differ for these two schemes. For indirect-drive the ablated material per time is approximately given by

$$\dot{m}\ (\text{g/cm}^{-2}\text{s}^{-1}) \cong 3 \times 10^5 T^3 = 10^7 I_{15}^{3/4}$$

[g/cm^3] and I_{15} the flux in units of 10^{15} W/cm^2.

In the direct-drive case the dependencies are the following:

$$\dot{m} \text{ (g cm}^{-2}\text{s}^{-1}) = 2.6 \times 10^5 \left(\frac{I_{15}}{\lambda^4}\right)^{1/3}$$

$$P(\text{Mbar}) \cong 40 \left(\frac{I}{\lambda}\right)^{2/3}$$

Using Eq. 5.11 this is equivalent of implosion velocities v_{imp}^{direct} and $v_{imp}^{indirect}$ given by

$$v_{imp}^{direct} \cong 1.5 \times 10^8 (I\lambda^2)^{1/3} \ln \left(\frac{m_0}{m}\right) \qquad (5.16)$$

$$v_{imp}^{indirect} \cong 1.8 \times 10^7 I_{15}^{1/8} \ln \left(\frac{m_0}{m}\right). \qquad (5.17)$$

The absorbed laser energy goes into both the ablation of material from the outside of the capsule as well as the acceleration of the material toward the center. Therefore only a part of the total energy PdV applied to the implosion can be used for the compression of the fuel. The hydrodynamic efficiency η_h reflects this fact and is defined as the ratio between the energy E_{imp} usable for the implosion and the energy E_a going into the ablation,

$$\eta_h = \frac{E_{imp}}{E_a} = \frac{\frac{1}{2}mv_{imp}^2}{E_a}. \qquad (5.18)$$

In the reference frame of Fig. 5.4, it follows from energy conservation that

$$-\frac{v_{exp}^2}{2}\frac{dm}{dt} = \frac{dE_a}{dt},$$

which means for the ablation energy E_a

$$E_a = -\frac{v_{exp}^2}{2}(m - m_0).$$

Substituting this expression into Eq. 5.18, it follows that the hydrodynamic efficiency is given by

$$\eta_h = \frac{\frac{1}{2}mv_{imp}^2}{\frac{1}{2}(m - m_0)v_{exp}^2} = \frac{m}{m - m_0}\frac{v_{imp}^2}{v_{exp}^2}.$$

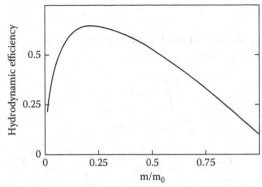

Figure 5.5. Hydrodynamic efficiency η as a function of the mass ratio m/m_0 according to Eq. 5.20.

Using Eq. 5.12 the hydrodynamic efficiency can either be expressed in terms of the velocities

$$\eta_h = \left(\frac{v_{imp}}{v_{exp}}\right)^2 \frac{1}{\exp(v_{imp}/v_{exp} - 1)}$$

or the masses

$$\eta_h = \frac{m}{m - m_0} \ln^2 \frac{m}{m_0}. \tag{5.19}$$

Equation 5.19 is often expressed as function of the rest mass ratio $x = m/m_0$, which then reduces to:

$$\eta_h = \frac{x \ln^2 x}{1 - x} \tag{5.20}$$

In Fig. 5.5 the hydrodynamic efficiency is plotted as a function of the rest. It can be seen that even under ideal conditions the hydrodynamic efficiency is less than 1 for any given m/m_0. The maximum achievable efficiency is about 70% for constant v_{exp}. However, there is substantial thermal energy in the ablating material, so that, according to Hatchett and Rosen (1993), the overall efficiency is in fact reduced to

$$\eta_h = \frac{4(\gamma - 1)}{5\gamma - 3} \frac{T_2}{T_x} \frac{x \ln^2 x}{1 - x}, \tag{5.21}$$

where $x = m/m_0$, T_2 is the temperature at the ablation, and T_x is the temperature at low density so that $T_2/T_x \sim 0.5$. So for a monoatomic gas

implosion velocity $_{imp}$ at half of its initial capsule radius . Integrating the rocket equation in time gives

$$\frac{R}{2} = \int_0^{t_1} v \, dt$$

or, substituting Eq. 5.10,

$$\frac{R}{2} = -v_{exp} \int_0^{t_1} \ln x \, dt.$$

Substituting $dt = dm/\dot{m} = dx/\dot{x}$ and equivalently $x_1 = m_1/m_0$ relates R to the fraction of ablated mass and one obtains

$$\frac{R}{2} = -\frac{v_{exp}}{\dot{x}} \int_0^{x_1} \ln x \, dx = -\frac{v_{exp}}{\dot{x}} [x_1 \ln x_1 - x_1 + 1]$$

$$= -\frac{v_{exp}}{\dot{x}} \left[\frac{m_1}{m_0} \ln \frac{m_1}{m_0} - \frac{m_1}{m_0} + 1 \right]$$

with $\dot{x} = v_{abl}/\Delta R$, it follows for m_1/m_0

$$\ln\left(\frac{m_1}{m_0}\right) = \ln\left(1 - \frac{v_{abl}t_1}{\Delta R}\right) \equiv \frac{v_{imp}}{\sqrt{T}}. \qquad (5.22)$$

The aspect ratio is therefore related to the implosion velocity via

$$\frac{R}{\Delta R} = \frac{2v_{exp}}{v_{abl}} \left[1 - \left(1 + \frac{v_{imp}}{\sqrt{T}}\right) \exp\left(-\frac{v_{imp}}{\sqrt{T}}\right) \right]. \qquad (5.23)$$

For direct-drive the implosion velocity v_{imp} is usually smaller than v_{exp} (i.e., $1 < v_{imp}/\sqrt{T} < 4$). In this case Eq. 5.23 reduces to

$$R/\Delta R \simeq 2v_{exp}v_{abl} \ln(m_1/m_0) = \frac{v_{imp}^2}{v_{abl}v_{exp}} = \frac{\rho v_{imp}^2}{P_a} \qquad (5.24)$$

which means that the implosion velocity is given by

$$v_{imp} \sim \frac{1}{2} \frac{R}{\Delta R} v_{abl}. \qquad (5.25)$$

This equation again suggests that a thin shell would be preferable. However, as mentioned before, that hydrodynamic instabilities tend to make the

$$\frac{v_{imp}(\text{cm s}^{})}{4 \times 10^7} \simeq \beta^{3/5} \left(\frac{T(\text{eV})}{300} \right) \qquad \text{for} \qquad \frac{R}{\Delta R} \sim 30, \qquad (5.26)$$

which relates the implosion velocity to the radiation temperature.

Another important parameter for the target design is the ablation velocity v_{abl}. Because of the requirement of a compression close to Fermi degeneracy, it can be assumed that $P = 2\beta\rho^{5/3}$, and therefore

$$v_{abl} = \frac{\text{d}}{\text{d}t}\Delta R = \frac{\dot{m}}{\rho} = 0.017\beta^{3/5}T^{0.9} \qquad (5.27)$$

where β is a constant describing the deviation from the ideal Fermi gas.

In the above calculations no distinction between the fuel material and ablator material was made, both of which are present in an ICF target. If the driver energy is continously injected between two different materials the instantaneous conservation of momentum flux gives $mv_{imp}^2 = Mv_{exp}^2$. The driver intensity is converted into the energy flow $mv_{imp}^3/2 + Mv_{exp}^3/2$; therefore, the ratio of the energy fluxes in both directions are given by

$$\frac{\frac{1}{2}mv_{imp}^3}{\frac{1}{2}Mv_{exp}^3} = \left(\frac{M}{m} \right)^{1/2}.$$

This equation shows that the energy preferably goes to the lighter element, which is the reason for placing heavy material at the outside of the target.

How does the implosion velocity depend on the target construction in the rocket model? If a shell of in-flight thickness ΔR and density ρ_{shell} is uniformly accelerated with an acceleration a from its initial radius R_0 to $R_0/2$ by an ablation pressure p_a, its velocity is

$$v_{sh}^2 = 2aR_0/2 = \frac{p_a}{\rho_{sh}\Delta R}R_0.$$

The shell then compresses the fuel to a final pressure p_f given approximately by $\rho_{sh}v_{sh}^2$ (Kilkenny et al., 1994)

$$p_f \sim \rho v_{sh}^2 = p_a\frac{R_0}{\Delta R}.$$

One can perform a similar calculation to obtain the implosion velocity (Rosen and Lindl, 1983; Max et al., 1980).

that for material containing density perturbations or inhomogeneties the different regions will move at different velocities. As in most solids the sound velocity increases with pressure,

$$dc_s/dp > 0. \tag{5.28}$$

High-pressure disturbances catch up with low-pressure disturbances, so that the initial density profile $\rho(t = 0)$ as well as the pressure $p(t = 0)$ become distorted at later times. in this way a compression wave is steepened into a shock front.

In a real plasma, viscosity and thermal transport counteract steepening of the wave front. These dissipative forces eventually outbalance the steepening effect because of the density-dependent sound velocity. The wave profile remains unchanged and a steady shock wave proceeds more — or — less unaltered.

Under which conditions is a shock wave stable? There are two possible locations for disturbances: either in front of or behind the shock wave. In the first case, the disturbance in front of the shock has to move slower than the shock wave, otherwise it would create a shock front itself. In the second case, the disturbance has to be fast enough, otherwise the shock will decay. The condition for a stable shock is therefore

$$c_s + v_d \geq v_s > c_{sd}. \tag{5.29}$$

Using Eq. 5.29, the relations $c_{sd}^2 = (\partial p / \partial \rho)_{sd} = V(\partial p / \partial V)_{sd}$ and $v_s^2 = V_0(p_1 - p_0)/(V_0 - V_1)$ this condition becomes:

$$-\left(\frac{\partial p}{\partial V}\right)_{sd} < \frac{p_1 - p_0}{V_0 - V_1} < \left(\frac{\partial p}{\partial V}\right)_H. \tag{5.30}$$

In other words $(p_1 - p_0)/(V_0 - V_1)$ must be steeper than the slope of the isentrope of the initial state but less steep than the Hugoniot at the final state.

5.5 Compression Phase

To model the compression phase one can start using the equivalent of equation (4.4) in spherical geometry given by Yabe (1993)

$$\frac{d^2 R}{dt^2} = -\frac{4\pi R^2}{M} p_1.$$

If a homogeneous isentropic compression is assumed, the thickness ΔR of the fuel layer decreases at the same rate as the radius R. From mass conservation it follows that the compression is approximately given by

$$\frac{\rho}{\rho_0} = \left(\frac{R_0}{R}\right)^3 \tag{5.32}$$

$$\frac{p}{p_0} = \left(\frac{R_0}{R}\right)^{3\gamma}. \tag{5.33}$$

Because the shell pressure equals the applied pressure, combining Eqs. 5.31 and 5.33 leads to

$$\frac{d^2R}{dt^2} = -\frac{p_0}{\rho_0\Delta R_0}\left(\frac{R_0}{R}\right)^{3\gamma-2}.$$

In the case of an ideal gas ($\gamma = 5/3$) the solution of this differential equation is

$$R = R_0\left(1 - \chi t\right)^{1/2}$$

where $\chi = \sqrt{4p_0/R_0\Delta R\rho_0}$. Using Eq. 4.15 it then follows that

$$p = p_0\left(\frac{1}{1-\chi t}\right)^{3\gamma/2}.$$

This equation shows that the pressure should have a certain time-dependence — provided by a tailored pulse to achieve the required compression.

5.6 Spherically Convergent Shock Waves

As we saw in Section 3.7, single planar shock waves can only compress the plasma by a factor of 4. It was suggested there that the key to achieving the required high density in ICF targets is the use of multiple shocks combined with the spherical converging geometry.

As long ago as 1942, Guderley found a self-similar solution of the Euler equation for spherically converging shock waves. He showed that for spherically convergent shocks a compression by a factor of 33 is possible. His solution shows that initially the spherical shock leads to a density increase by a factor of 4 just like for a plane shock. However, in the spherically symmetric case this is followed by an isentropic compression

energy of (Brueckner and Jorna, 1974)

$$E_{driver} = 1.6 \frac{M_g^3}{\epsilon_D^4} \eta^2 \quad [\text{MJ}], \tag{5.34}$$

would be required. Here where M_g is the fusion gain, ϵ_D is the driver coupling efficiency and η the desired compression. More detailed studies showed that the required driver energy using a single shock to compress the pellet to breakeven conditions would be 500 MJ. This is far more than any driver could provide nowadays.

As just mentioned, it requires much less energy to perform an isentropic compression by a succession of multiple shocks, a process we now look at in more detail.

5.7   Isentropic Compression

If the DT fuel stays close to being Fermi degenerate during the compression, the compression energy is small compared with the ignition energy for the same amount of fuel. Every Fermi particle occupies a phase space of volume h^3, so N particles in a volume V therefore fill a phase space

$$Nh^3 = \sum_s \int_V \int_{p_f} d^3x d^3p.$$

The spin of the particles is $s = \frac{1}{2}$, so the sum over all spin states is 2. If p_{fermi} is the momentum of the highest energy particles, it follows that

$$Nh^3 = (2s+1)V \frac{4\pi}{3} p_{fermi}.$$

Expressed in terms of the Fermi energy, $\epsilon_f = p_f^2/2m$, this reads

$$\epsilon_f = \frac{\hbar^2}{2m_e} \left(\frac{6\pi^2}{2s+1} \frac{N}{V} \right)^{2/3}. \tag{5.35}$$

This means that $\epsilon_f(\text{eV}) = 14\rho^{2/3}(\text{g/cm}^3)$. Because the average energy per particle is 0.6 ϵ_f, the specific energy ϵ_{DT} of DT in J/g is related to the density the following way

$$\epsilon_{DT}(\text{J/g}) = 3 \cdot 10^5 \rho^{2/3}(\text{g/cm}^3). \tag{5.36}$$

$$(\text{J/g}) = 3 \quad \rho \quad (\text{g/cm}) \left(\frac{}{\rho^{4/3}} \right) \qquad (5.37)$$

This equation shows that temperature effects can be neglected as long as $0.02T_e^2(\text{eV}) \ll \rho^{4/3}$. For DT at 1 g/cm^3 the temperature effects do not play a role as long as the temperature stays below a few eV, but for higher temperatures these effects have to be considered. In the ICF context, ion contributions and molecular effects play role as well, which have not been included here.

5.8 Multiple Shocks

For an efficient compression it is necessary to produce a sequence of shock waves that follow the adiabatic compression curve for the fuel as closely as possible. This can be achieved by shaping the laser pulse in such a way that the pressure on the target surface gradually increases, so that each shock generated rises in strength.

When the shocks reach the center their kinetic energy is converted into thermal energy, leading to a temperature increase: they are then reflected and propagate outward. However, this leads to an additional requirement for successful compression — these shocks have to be timed in such a way that they do not overtake each other but all arrive at the center of the capsule simultaneously. Obviously the final temperature in the center is determined by the combined strength of the shocks and preheat has to be avoided.

A simple model for a perfectly isentropic compression along the adiabat can give an estimate of the energy requirement for this process. In a simple piston approach the work performed between the states before and after the compression, indicated by the subscripts 0 and 1, respectively, is given by

$$W_{0 \rightarrow 1} = \int_0^1 p \, dV = const.$$

Assuming an equation of state of the form $pV^\gamma = $ constant, the performed work is

$$W_{0 \rightarrow 1} = C \int_0^1 V^{-\gamma} dV, \qquad (5.38)$$

where C is a constant. The states before and after the compression are related via

$$p_0 V_0^\gamma = p_1 V_1^\gamma$$

$$W_{0 \to 1} = \frac{p_1 V_1 - p_0 V_0}{1 - \gamma} = \frac{nk}{1 - \gamma} T_0 \left[\left(\frac{V_0}{V_1} \right) - 1 \right]. \qquad (5.39)$$

Equation 5.39 shows that only a relatively small amount of the required driver energy actually goes into the compressed fuel. Most of the energy is used to produce the pressure to drive the compression. For example, compressing 1 mg of DT fuel with an initial temperature of 1 eV to 1000 times its liquid density requires only of the order of 6 kJ — far less than the required driver energy.

In Section 5.2 we saw that the energy requirements for compression is reduced significantly by using not a solid target but targets that contain just an outer shell of DT fuel. For such shell targets Kidder (1976a,b) developed a theory based on self-similar solutions. The pressure profile he obtained depends on the relative size of the capsule radius R_0 to the shell thickness ΔR_0 and has the following form

$$\frac{p(t)}{p_0} = P_{profile} = \left[1 + \left(\frac{t}{t_c} \right)^2 \right]^{-5/2}, \qquad (5.40)$$

where $t_c = (R_0^2 - \Delta R_0^2)/3c_s^2$. This would imply that such a compression could result in arbitrarily large densities.

The reason why this is not so in reality is that the required compression and fusion temperature have to be achieved simultaneously just before ignition. This means that in the last stages before ignition the pressure profile has to deviate from that of an isentropic compression. In this final phase Kidder's calculations suggest a pressure profile of the form

$$P_{end-phase} = P_{profile} \exp \left(5 t_a \frac{t - t_a}{t_c^2 \ t^2} \right) \qquad (5.41)$$

where t_a is usually 0.9 t_c. This would mean that roughly a sixth of the original capsule mass can be compressed by a factor of 10,000.

In the next section we demonstrate what happens if the compression is successful, the conditions for ignition are fulfilled and ignition actually begins.

5.9 Burn

The energy gain in ICF fusion obviously depends on the amount of fuel burned in the process. However, whatever target configuration is chosen,

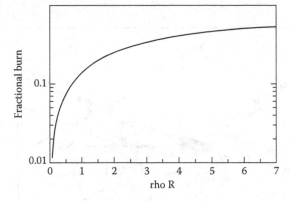

Figure 5.6. Fractional burn as a function of ρR as given by Eq. 5.47

it will never be possible to burn the entire fuel. The fractional burn f_b can be derived (Meyer-ter-Vehn, 1982) from the nuclear reaction rate given by

$$\frac{dn}{dt} = -n_D n_T \langle \sigma v \rangle \tag{5.42}$$

where n is the number of thermonuclear reaction per time unit, n_D and n_T are the ion number densities of deuterium and tritium and $\langle \sigma v \rangle$ the averaged reaction cross section for particles with a Maxwellian velocity distribution. If the ion density of deuterium and tritium are the same, it follows that

$$n_D = n_T = \frac{n_0}{2} - n,$$

where n_0 is the initial total number density. Introducing the fractional burn as the ratio of the number density of reaction products to the initial number density of the deuterium (or tritium), or $f_b = n/n_{DT} = n/(n_0/2)$, the number density n can be expressed as $n = n_0 f_b/2$. Equation 5.42 can then be rewritten as follows

$$\frac{n_0}{2} \frac{df_b}{dt} = \left(\frac{n_0}{2} - \frac{n_0 f_b}{2} \right)^2$$

or

$$\frac{df_b}{dt} = \frac{n_0}{2}(1 - f_b)\langle \sigma v \rangle. \tag{5.43}$$

Assuming that $\langle \sigma v \rangle$ is constant during burn time τ_b, it follows that

$$\frac{f_b}{1 - f_b} = \frac{n_0 \tau_b}{2} \langle \sigma v \rangle.$$

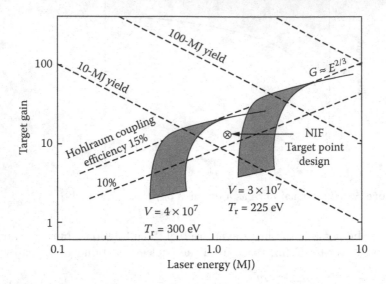

Figure 5.7. Target gain as function of the laser energy for an indirect drive target assuming a hohlraum coupling efficiency of 10% and 15%, respectively. For given implosion velocities ($v = 4 \times 10^7$ cm/s and $v = 3 \times 10^7$ cm/s) the expected gain region for optimistic and pessimistic assumptions is shown. Reprinted with permission from Lindl (1995) ©1995, American Institute of Physics.

This can be rewritten as

$$f_b = \frac{n_0 \tau_b \langle \sigma v \rangle / 2}{1 + (n_0 \tau_b \langle \sigma v \rangle / 2)}. \qquad (5.44)$$

The burn time and the sound speed are approximately related by $\tau_B = r/3c_s$ and for Eq. 5.44 it follows

$$f_b = \frac{n_0 r \langle \sigma v \rangle / (6 c_s)}{1 + (n_0 r \langle \sigma v \rangle) / (6 c_s)}.$$

Using the mass density instead of ion density, we have finally

$$f_b = \frac{\rho R}{\rho R + \psi(T_i)} \qquad (5.45)$$

where $\psi(T_i) \sim C_s / \langle \sigma v \rangle$. The reaction rate $\langle \sigma v \rangle$ depends strongly on the temperature. Hiverly (1977) approximated this by the following formula

$$\langle \sigma v \rangle = \exp(a_1 / T_i^r + a_2 + a_3 T_i + a_4 T_i^2 + a_5 T_i^3 + a_6 T_i^4) \qquad (5.46)$$

$$a_3 \qquad 7.1013427 \times 10$$
$$a_4 = 1.9375451 \times 10^{-4}$$
$$a_5 = 4.9246592 \times 10^{-6}$$
$$a_6 = -3.9836572 \times 10^{-8}$$

$r = 0.2935$ and the temperature is given in keV. For a temperature of 10 keV, ψ is found to be about 19 g/cm^3 and for 20 keV about 6.8 g/cm^3.

An often-used approximation of formula 5.45 is

$$f_b = \frac{\rho R}{\rho R + 6(g/cm^2)}, \tag{5.47}$$

which is valid for DT between 20 and 40 eV. This simple formula agrees very well with more detailed numerical simulations (Nuckolls, 1994) of the burn process of most ICF targets. For example, it shows that a 33% fractional burn requires a $\rho r = 3\text{g/cm}^2$.

The aim in inertial fusion is to maximize the fusion energy output for a given driver energy input, so it is important that the burn phase proceeds efficiently as well. Here the goal is to have the right fuel densities in the hot spot region so that the fusion products (neutrons and α-particles) are stopped in this area and deposit their energy. In this way the temperature increases via self-heating, and more fusion reactions can take place.

In a first approximation the heating by the neutrons can be neglected when considering the self-heating in the hot spot area, because their range in 10 keV plasmas is about 20 times that of the α-particles so they will mainly deposit their energy outside the hot-spot area.

For efficient self-heating ρr must therefore be such that the α-particles are stopped within a fraction of the hot-spot region,

$$\rho R \gg \rho \lambda_\alpha, \tag{5.48}$$

where λ_α is the mean-free path of the α-particles. As soon as the burn process has started the energy deposition of the α-particles will quickly heat the core to 20–80 keV (Henderson, 1974; Beynon and Constantine, 1977; Mason and Morse, 1975).

The α-particle energy deposition their energy in the hot spot area can be described by the Bethe stopping formula (see Section 10.2)

$$\left(\frac{dE}{dx}\right)_{Bethe} = \frac{4\pi N_0 Z_{eff}^2 \rho_o^{stop} e^4 Z_{stop}}{m_e c^2 \beta^2 A_{stop}} \left[\ln \frac{2m_e c^2 \beta^2 \gamma^2}{I_{av}} - \beta^2 \sum_i \frac{c_i}{Z_{stop}} - \frac{\delta}{2} \right],$$

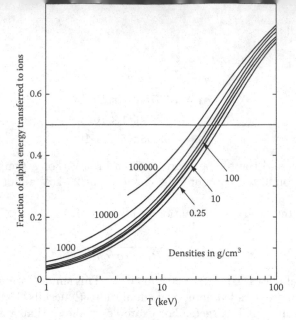

Figure 5.8. The relation between the fraction of the energy of the α-particle that is transfered to the ions for different temperatures.

where the index $_{stop}$ stands for the to-be-stopped particle type, and I_{av} the average ionization potential.

For the stopping of α-particles in a plasma, Fraley et al. (1974) found:

$$
\left(\frac{dE_\alpha}{dx}\right) = -\,26.9 \left(\frac{\rho}{\rho_0}\right) \frac{E_\alpha^{1/2}}{T_e^{1/2}} \left(1 + 0.168 \ln\left[T_e \left(\frac{\rho}{\rho_0}\right)^{1/2}\right]\right)
$$
$$
-\,0.05 \left(\frac{\rho}{\rho_0}\right) \frac{1}{E_\alpha} \left(1 + 0.075 \ln\left[T_e^{1/2} \left(\frac{\rho}{\rho_0}\right)^{1/2} E_\alpha^{1/2}\right]\right),
$$

where E_α is in units of 3.5 MeV and ρ_0 in units of the solid density of DT (0.25 g/cm^2). The first term of the energy loss is due to the interaction with the electrons and the second term due to the interaction with the ions. For the fusion process the energy that is transfered to the ions is the relevant one. The fraction f_{ion} of the energy deposited in the ions and not the electrons is given by

$$
f_{ion} = \left(1 + \frac{32}{T_e}\right)^{-1}.
$$

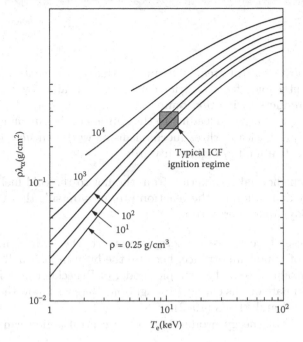

Figure 5.9. The α-particle range $\rho\lambda_\alpha$ as a function of the electron temperature T_e. Reprinted with permission from Lindl (1995) ©1995, American Institute of Physics.

Figure 5.8 shows the fraction f_{ion} of the α-energy that goes to the ions as a function of the temperature for different densities ρ. For low temperatures most of the deposited energy goes to the electrons. At solid density, low temperature means below approximately 32 keV and at $T_e = 10$ keV only about 25% of the energy of the α-particles will be deposited in the ions and therefore used to heat the central core region. The range of α-particles in solid-DT can be approximated by (Frayley *et al.*, 1974)

$$\rho\lambda_\alpha(\text{g/cm}^2) = \frac{1.5 \cdot 10^{-2}T_e^{5/4}}{1 + 8.2 \cdot 10^{-3}T_e^{5/4}} \tag{5.49}$$

with the electron temperature T_e in keV. Figure 5.8 shows the range of the α-particles as a function of the temperature. The density of the hot-spot area lies in the range of 10–100 g/cm^3 with a temperature of about 10 keV. The ratio of the α-particle range to the radius R of the compressed fuel area can be approximated by Frayley *et al.* (1974)

$$\frac{\lambda_\alpha}{R} \sim \frac{1.9}{1 + 122/T_e^{5/4}} \frac{1}{\rho R}. \tag{5.50}$$

work heats the central area. This means that as the central spark burns, the adjacent cooler fuel material can become heated by the outflowing reaction products to ignition temperatures. In this way a spherical burn wave propagates outward leading to the ignition of the surrounding plasma.

There are basically three mechanisms at work transfering the energy from the central ignited region outward:

- electron thermal conduction from the hot to the cold fuel regions
- energy deposition by the reaction products outside the central region
- hydrodynamic energy transfer

Of these three types of energy transport, the hydrodynamic energy transfer is of minor importance, because the burn front usually propagates with supersonic speed. The simple model of Brueckner and Jorna (1974) gives an estimate of this energy propagation. Here we follow the description of Duderstadt and Moses (1982).

The rate of energy production in the central region can be approximated by

$$\frac{dE_{fusion}}{dt} = \frac{4\pi r^3}{3} \langle v\sigma \rangle \frac{n_0^2}{4} W_\alpha,$$

where a uniform density n_0 is assumed, W_α is the α-particle energy deposition and neutron energy production is ignored. This corresponds to the rate of change in the internal energy in the expanding region

$$\frac{dE_{int}}{dt} = \frac{d}{dt}\left(4/3r^3 n_0 k_B T_0\right) = 4/3r^3 n_0 k_B \frac{dT_0}{dt} + 4n_0 k_B T_0 r^2 \frac{dr}{dt}.$$

as the burn region extends. Here T_0 is the temperature of the burning region. Combining these two equations, and using $dE_{fusion}/dt = dE_{int}/dt$, one obtains for the change of the radius of the burn region

$$\frac{dr}{dt} = \frac{n_0 \langle v\sigma \rangle W_\alpha r}{12 k_B T_0} - \frac{r}{3T_0}\frac{dT_0}{dt}. \tag{5.51}$$

The temperature in the burn region, T_0, will increase until the α-particles are able to escape into the surrounding colder fuel, then the temperature in the burn region will adjust itself that $r \sim \lambda_\alpha$, so that for $T_0 > 40$ keV

$$r \sim \lambda_\alpha = \lambda_0 \frac{T^{3/2}}{n_0}.$$

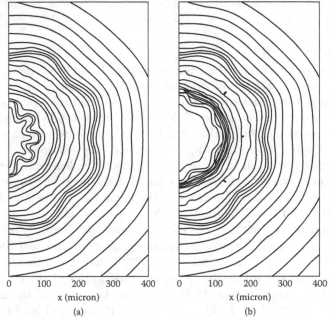

Figure 5.10. Comparison between the density contours at the maximum compression of a high gain implosion a) without and b) with including the α-heating effect (Takabe and Ishii, 1993).

It follows that

$$\frac{1}{T_0}\frac{dT_0}{dt} \sim \frac{2}{3R}\frac{dr}{dt}.$$

Now $v_{burn} = dr/dt$ the speed of the burn front. Comparing it with the speed of sound $c_s = v_0 T^{1/2}$, we obtain

$$\frac{v_{burn}}{c_s} \sim 3\langle v\sigma\rangle W_\alpha \frac{\lambda_0}{44v_0}.$$

It follows that in the cold fuel material surrounding the hot spot the propagation of the burn front is supersonic, because $v_{burn}/c_s > 2$, if $kT_0 > 15$ keV.

Obviously the simple model described above does not take into account all processes going on in the burn phase. Proper investigations need much more detailed numerical simulations of the hydrodynamic processes. In addition one needs to take into account processes at the atomic level. Another point to consider is the following: as deuterium and tritium ions move through the fuel, they are slowed down. The primary process responsible for this is binary electron collisions (Cable, 1995). Therefore the

and tritium will still have their initial energy. In addition there are processes that counteract the heating of the hot-spot area; namely, electron conduction to the colder surrounding plasma and radiative processes.

Apart from increasing the effciency of the burn, self-heating has an additional positive effect, in that it helps to suppress hydrodynamic instabilities originating from the compression phase. As mentioned before and discussed in more detail in Chapter 6, a major problem in achieving the compression of the plasma to the required densities is the occurrence of instability induced nonuniformities in the compression. Takabe and Ishii (1993) carried out two-dimensional hydrodynamic calculations of high-gain implosions with and without the effect of α-heating (see Fig. 5.10). The nonuniformities manifest themselves as the spiky structures near the central region in Fig. 5.10a. Takabe and Ishii (1993) found that after the ignition and self-heating has taken place, the nonuniformities of the compression introduced by instabilities in the plasma near the spark and the main fuel interface is smoothed by the effect of the α-particles (see Fig. 5.10b).

Chapter 6

Rayleigh-Taylor Instabilities

In previous chapters it has been stressed that a key point in achieving fusion is a homogenous compression, which means aiming for a perfectly spherically implosion. In reality this ideal is never reached, which has a number of consequences:

- The conversion of kinetic energy from the implosion into internal energy of the fuel is imperfect, reducing the maximum compression.
- Severe perturbance of a spherically symmetric implosion can lead to small scale turbulences and even to break-up of the shell.
- The hot-spot area is increased or has a larger surface because of the uneven structure, which in turn reduces the achievable temperature and can cause the α-particles created to escape the hot-spot area earlier, thus lowering the self-heating (see Fig. 6.1).

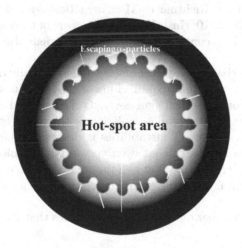

Figure 6.1. Schematic picture of the reduced self-heating of the hot-spot area from prematurely escaping α-particles.

126

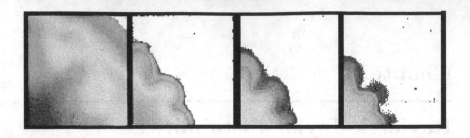

Figure 6.2. Growth of Rayleigh–Taylor instabilities during implosion. ©ENEA, Italy.

The maximum acceptable defect ΔR_f of the final radius R_f is about 33% (Lindl, 1995; Andre *et al.*, 2003) (i.e., $\Delta R_f/R_f < 1/3$); if it is higher, the fusion process is not successful. The convergence ratio $C = R_i/R_f$ describes the ratio of the initial to the final radius of the capsule, and a typical value is $C = 30$. With

$$\Delta R_f/R_f = C\Delta v/v < 1/3,$$

it follows therefore that a implosion velocity uniformity better then 1% is required for a successful inertial confinement fusion (ICF) implosion.

The main hindrance to a spherically symmetric compression are instabilities, of which the Rayleigh–Taylor (RT) instabilities (Taylor, 1950)) are the dominant ones in ICF and mainly described in this chapter. Apart from RT instabilities, Richtmeyer-Meshkov (Richtmyer, 1960) and Kelvin-Helmholtz (Kelvin, 1910) instabilities can occur, but because they are less important in the ICF context they will only be briefly discussed in Section 6.6.

Plasma disturbances can either decay and eventually disappear so that the plasma returns to a stable equilibrium; or they start to grow, amplifying any deviations from spherical symmetry. Obviously if the latter happens, there is a risk to the successful completion of the compression phase. The devastating effect of the RT-instabilities is due to the fact that they initially grow exponentially, so that even very small, seemingly insignificant disturbances can reach a size that can threaten the whole compression (see Fig. 6.2). The aim is therefore to minimize these asymmetries in the first place.

There are a number of causes of disturbances that lead to instabilities, principally

- nonuniform radiation and
- the quality of the manufactured target.

within the single beams themselves that can seed instabilities. Another source for instabilities can be the timing of the beams. If this is not perfect or the pulse shaping not exactly synchronized, this causes disturbances as well. Another example is the interference pattern in a focused laser spot, which can imprint disturbances on an initially smooth surface irradiated by the laser.

Direct- and indirect-drive targets are affected by RT instabilities to different degrees. This is because direct-drive targets are irradiated by a limited number of beams whereas the x-ray illumination in indirect-drive targets leads to a much more homogeneous radiation field. However, even in indirect-drive schemes, avoiding disturbances from irradiation inhomogeneities is vital.

It would be premature to conclude that indirect-drive targets are generally less susceptible to instabilities. Indirect-drive targets create a different set of problems. The plasma electron density n_e in indirect-drive hohlraum targets is typically a few percent of the critical density n_c, so stimulated Raman scattering and stimulated Brillouin scattering are a problem (see Chapter 4). These have very large predicted linear growth rates because of the long scale length of these nearly uniform plasmas. However, as mentioned earlier, the understanding of the nonlinear saturation mechanisms is still relatively poor.

Returning to the problem of RT instabilities, the targets themselves can cause the instabilities for the simple reason that they are prone to technical imperfections during manufacturing. The quality of the surface finish is crucial, because even small machining marks from the production process or the crystalline structure of the material can seed disturbances. Slight variations in the thickness of the different target layers can also lead to inhomogeneities in the resulting plasma. In Chapter 8 we will discuss the quality requirements for the target manufacturing in more detail.

In summary, it can be said that all these causes of perturbations in the implosion have to be minimized to avoid RT instabilities as far as possible, otherwise they may have disastrous consequences for the compression. To understand why this is so, we now look at the physics of RT instabilities from first principles.

6.1 Basic Concept

In its original form the RT-instability describes an effect that occurs in a system of two incompressible fluids. A heavy fluid lying on top of a lighter

lower density (hot plasma) on a higher density (cold plasma) and instead of gravity, which in this case is negligible, the ablation pressure acts as transmitting force. Interestingly, the same phenomenon occurs in supernovae implosions, as well (Kane *et al.*, 2000).

To understand the nature of RT instabilities, let us start with very simple considerations. In the previous chapters we noted that for various reasons the implosion will always exhibit some kind of nonuniformity. For a constant acceleration a, the implosion time of a target with initial radius R_0 can be estimated from the relation

$$R_0 = \frac{1}{2}at^2. \tag{6.1}$$

In the presence of a nonuniformity in the acceleration Δa, different parts of the plasma will move at different speeds, eventually resulting in a interface perturbation

$$R_0 + R_{per} = \frac{1}{2}(a + \Delta a)t^2$$

after this time. Elimination of t gives

$$R_0 + R_{per} = (a + \Delta a)\frac{R_0}{a},$$

which simplifies to

$$\frac{R_{per}}{R_0} = \frac{\Delta a}{a}.$$

So in this model the plasma perturbation depends linearly on the acceleration difference Δa. What would that mean for the compression?

Recall that the volume compression C depends on ratio of the initial to the final radius R_f of the target, $C = (R_0/R_f)^3$. Because the perturbation R_{per} cannot be larger than the final radius of the target — otherwise the target would just fly apart — the maximum compression is given by

$$C_{max} = \left(\frac{R_0}{R_{per}}\right)^3 = \left(\frac{a}{\Delta a}\right)^3. \tag{6.2}$$

This shows that the achievable compression depends strongly on the size of the perturbation — the smaller the nonuniformity the larger the possible compression. The above picture is of course oversimplified: even if

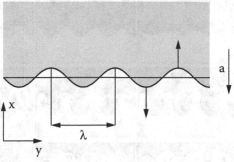

Figure 6.3. Schematic picture of RT instabilities.

the nonuniformity is very small in the interaction phase, it can still be amplified during the implosion, an effect that demands a more sophisticated treatment:

We start from the single fluid model given by Eqs. 3.8–3.10 reproduced here as

$$\frac{\partial \rho}{\partial t} + v_x \frac{\partial \rho}{\partial x} + v_y \frac{\partial \rho}{\partial y} = 0$$

$$\rho \left(\frac{\partial v_x}{\partial t} + v_x \frac{\partial v_x}{\partial x} \right) = -\frac{\partial P}{\partial x} - \rho a$$

$$\rho \left(\frac{\partial v_y}{\partial t} + v_y \frac{\partial v_y}{\partial y} \right) = -\frac{\partial P}{\partial y}$$

$$\frac{\partial P}{\partial t} + v_x \frac{\partial P}{\partial x} + v_y \frac{\partial P}{\partial y} - 0.$$

Combining the first and third equation, it follows for an incompressible fluid that $\nabla v = 0$. Imagine now a situation like in Fig. 6.3, where a wave of a certain wavelength λ disturbs the system in such a way that the following ansatz can be used

$$f = f_0(x) + f_1(x) \exp(iky + \gamma t), \qquad (6.3)$$

where f_0 is the equilibrium solution, f_1 the perturbation, and k the wave number $(2\pi/\lambda)$ of the instability. It follows for the equilibrium (denoted by the subscript 0)

$$\frac{\partial P_0}{\partial x} = -\rho_0 a.$$

With $v_0 = 0$ and $v_1 = (v_x, v_y)$ and the two spatial components for the momentum conservation written in separate equations, one obtains

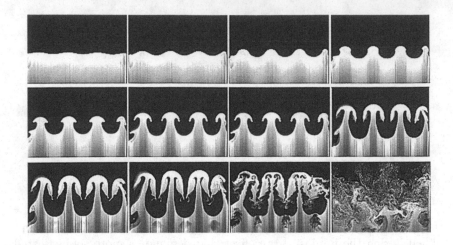

Figure 6.4. RT instabilities in a two-fluid system. This picture shows experimental results of accelerating a tank containing two fluids downwards at a rate greater than the earth gravitational acceleration. Reused with permission from J. T. Waddell, C. E. Niederhaus, and J. W. Jacobs, Physics of Fluids, 13, 1263 (2001). Copyright 2001, American Institute of Physics.

$$\gamma \rho + v_x \frac{\partial \rho_0}{\partial x} = 0$$

$$\rho_0 \gamma v_x = -\frac{\partial P_1}{\partial x} - \rho_1 a$$

$$\rho_0 \gamma v_y = -ikP_1,$$

$$\gamma P_1 + v_x \frac{\partial P_0}{\partial x} = 0.$$

For the incompressibility condition one obtains:

$$0 = \frac{\partial v_x}{\partial x} + ikv_y$$

and it follows

$$v_y = \frac{i}{k} \frac{\partial v_x}{\partial x}.$$

Eliminating P_1 and ρ_1 from the momentum equations yields

$$\rho_0 \gamma v_x = \frac{1}{\gamma} \frac{\partial}{\partial x} \left(v_x \frac{\partial P_0}{\partial x} \right) + \frac{v_x}{\gamma} \frac{\partial \rho_0}{\partial x} a$$

$$\rho_0 \gamma v_y = \frac{ikv_x}{\gamma} \frac{\partial P_0}{\partial x} = -\frac{ikv_x}{\gamma} \rho_0 a.$$

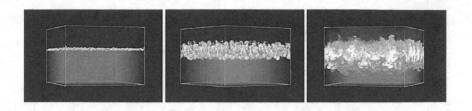

Figure 6.5. Three-dimensional RT instabilities, ©LLNL[1].

Combining the v_y equation with the incompressibility condition it follows

$$\frac{\partial v_x}{\partial x} = -\frac{k^2}{\gamma^2}av_x.$$

Integration gives

$$v_x(x) = w_0 \exp\left(-\frac{k^2}{\gamma^2}ax\right).$$

Substitute $\partial v_x/\partial x$ into the momentum equation, we finally obtain the dispersion relation

$$\gamma^4 = k^2 a^2$$

or

$$\gamma^2 = \pm ka.$$

The RT growth rate is then

$$\gamma = \sqrt{ka}. \tag{6.4}$$

A key parameter characterizing the potential damage of RT instabilities to the compression is the number of e-foldings, n_{max}, defined as

$$n_{max} = \int \gamma_{max} dt. \tag{6.5}$$

Assuming constant acceleration Eqs. 6.4 and 6.1 can be combined to obtain

$$n_{max} = \frac{1}{2}\sqrt{\frac{R}{\Delta R}}, \tag{6.6}$$

which links the RT instability directly to the aspect ratio of the target. This is why RT instabilities directly influence the target design. In Chapter 5 we

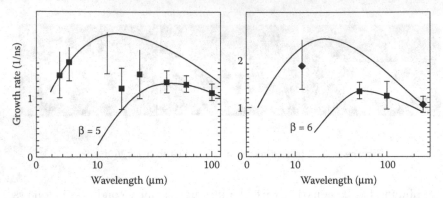

Figure 6.6. Experimental and theoretical growth rates of the RT instability (Azechi *et al.*, 2003).

saw that a large implosion velocity requires a large $R/\Delta R$, whereas Eq. 6.6 shows that RT instabilities set an *upper* limit to the aspect ratio. Current targets are designed for an aspect ratio $R/\Delta R \sim 30$, corresponding to just three e-foldings, or an amplification factor of 5.

In an ICF compression the interface between the two "fluids" will not be a sharp discontinuity as assumed above, but instead will have a continuous density gradient. This gradient can be very steep at places, but will usually vary continously over a distance K comparable to the perturbation length. By carrying out a similar calculation to the one just shown for a density profile of the form

$$\rho_0(z) = \rho_0 + \frac{1}{2}\Delta\rho\exp[+Kz], \text{ for } z < 0$$

$$= \rho_1 - \frac{1}{2}\Delta\rho\exp[-Kz], \text{ for } z < 0,$$

where $\Delta\rho = \rho_1 - \rho_0$, Lelevier *et al.* (1955) found that the growth rate is modified to

$$\gamma_{RT} = \sqrt{\frac{ka}{1 + kL}}. \tag{6.7}$$

where L is the density gradient length defined as $L = 1/K$. The effect of a continuous density gradient instead of a discontinuity is a reduction of the growth rate. For steep gradients $L \longrightarrow 1$ or perturbation wavelength much longer than L, it follows that result $\gamma = \sqrt{ka}$ is recovered. In the opposite limit of shallow gradients, i.e. $kL \gg 1$, the growth rate becomes independent of the wavelength, $\gamma \longrightarrow \sqrt{a/L}$.

6.2   RT in the Ablation Phase

In the ablation phase the energy is deposited in a narrow low-density region in the plasma, where a high-pressure is created directly next to the high-density layer that becomes accelerated inward. This is a similar situation to the one discussed in Section 6.1, the only difference being that there is an additional flow of material across the ablation surface from the high-density region into the low-density plasma.

Early estimates (Nuckolls, 1972) of the RT instability growth rate γ include the effect of ablation with an expression of the form:

$$\gamma^2 = ka - k^2 \frac{P_a}{\rho} = ka[1 - k\Delta R].$$

It soon turned out that this considerably overestimates the stabilization effect, leading to the erroneous conclusion that a minimum driver energy of 1 kJ would have sufficed to achieve fusion. Nevertheless, a similar approach, but with the assumption that one of the fluids initially moves with a finite velocity (ablation velocity v_{abl}^*) is capable of describing the RT in the ablation phase quite well.

In 1974 Bodner formulated a simple model for the ablation situation with a discontinuity in the density, which shows that the growth rate of the instability is reduced below the classical value $\sqrt{ka}$ by mass ablation to

$$\gamma \sim \sqrt{ka} - kv_a,$$

where v_a is the flow velocity across the ablation front. However, he needed to introduce an ad hoc assumption to close the problem.

Because the RT growth rate is so critical for determining the required driver energy, many attempts have been made over the years to derive it analytically. Gamaly (1993) derived the following expression

$$\gamma_{RT} = \sqrt{\frac{ka}{1 + kL}} - kv_{abl}^*, \tag{6.8}$$

which includes both the ablation velocity and the continuous density gradient in the calculations. However, experimental results (Desselberger and Willi, 1993; Grun *et al.*, 1987) and numerical simulations (Tabak *et al.*, 1990; Gardner *et al.*, 1991) have indicated larger stabilization effects, such

$$\gamma \qquad\qquad abl, \qquad\qquad\qquad (6.9)$$

with α and $b = 3 - 4$ (obtained by fitting to numerical simulations). Sanz (1994) finds better agreement with recent experiments (Remington et al., 1993) for $b = 2$.

It should be noted that here the ablation velocity v_{abl}^* means the ablation velocity divided by the density at the ablation surface, whereas in Eq. 6.8 the ablation velocity denotes the final velocity.

Today the most widely used growth rate is a modification of Eq. 6.8, namely

$$\gamma_{RT} = \sqrt{\frac{ka}{1 + kL}} - \beta_{RT} k v_{ablation}. \qquad\qquad (6.10)$$

This is still only an analytical fit to numerical results, where β_{RT} is a constant between 1 and 3. $\beta_{RT} \sim 1$ corresponds to an indirect-drive scenario whereas $\beta_{RT} = 3$ to the direct-drive case. The term $-\beta_{RT} K v_{ablation}$ describes the stabilizing effect of ablation, but β_{RT} is smaller for indirect-drive than direct-drive, seemingly in contradicting what was said before about indirect-drive being less susceptible to RT instabilities.

The reason why indirect-drive is still less sensitive is that the ablation velocity is much higher than in direct-drive. For a typical laser intensity of 10^{15} W/cm^2, the ablation velocity is about 10 times higher in indirect-drive than direct-drive implosions. So overall, ablation stabilizes against RT about a factor of 3 better in indirectly driven implosions than in directly driven ones.

The difference to the directly driven implosion is that the soft x-rays act like a very short-wavelength broadband laser. In this way they are able to penetrate further into the target and the energy is deposited over a larger portion of the shell.

The acceleration a and the wave number k are time-dependent and therefore the initial perturbation R_0^{per} is amplified to

$$R_{per} = R_0^{per} \exp(\alpha \int \gamma dt), \qquad\qquad (6.11)$$

where α describes the fact that the perturbed surface partly ablates and therefore stabilizes the instability. For ICF targets it is about 0.25 to 0.5 in the acceleration phase (Emery et al., 1982).

The maximum number of e-foldings can be calculated in both cases using Eqs. 6.5 and 6.10 (see Lindl, 1995). For the direct-drive case the

for a laser frequency of $\lambda = 1/3$ μm. For indirect-drive the maximum number of e-foldings is given by

$$n_{max}^{indirect} \sim \sqrt{\frac{kR}{1 + 0.2kR(\Delta R/R)}} - 0.8kR\frac{\Delta R}{R}. \qquad (6.13)$$

where kR is the Legendre polynomial mode number. Most of the shell mass is ablated in this case and therefore $1 - m/m_0 = 0.8$. It should be pointed out that in Eq. 6.13, ΔR is not the initial shell thickness but the average shell thickness, usually approximated by taking the value when the shell has been accelerated to about half its maximum velocity, which is at about a quarter of the initial radius. If in Eq. 6.13 the number of e-foldings is set to $n_{max}^{indirect} \sim 6$, then this implies that $R/\Delta R \sim 30$, the value for many current target designs quoted previously.

Equation 6.8 shows that the growth of short wavelengths is reduced more effectively than that of long wavelengths. In fact there actually exists a lower limit cut-off wavelength λ_{RT} given by

$$\frac{1}{\lambda_{RT}} = \left(\frac{\alpha}{b}\right)^2 \frac{a}{v_a^2} \qquad (6.14)$$

for which the growth is prohibited altogether, and below which modes are stable. This fact and the reduction of the growth rate by ablation is crucial for a successful compression in ICF.

So far our treatment of RT instabilities has assumed that the amplitude of the perturbation is small, allowing a linear analysis. However, the exponential growth implies that these perturbations will eventually become large, invalidating this assumption. When the development of the instability departs from the linear regime it is said to saturate. In the two-layer picture saturation is reached, when the displacement η_d from the interface between the hot and cold fuel no longer fulfills the condition

$$\frac{\eta_d}{\lambda} \ll \frac{1}{\sqrt{3}} \frac{1}{2\pi} \sim \frac{1}{10}. \qquad (6.15)$$

This means as soon as the amplitude of the perturbation exceeds about 10% of the wavelength, the growth rate decreases and is no longer exponential. At this stage the instability loses its sine-shape and a "bubble-and-spike" topology develops similar to those seen in Figs. 6.3 and 6.5. These structures no longer grow isolated from each other as in the linear stage, but start to influence each others' growth. This effect is called mode coupling, which can be most dramatically seen in the last picture of the sequence shown in Fig. 6.3.

the inside of the high-Z shell, or these which feed through from the outside of the shell.

Here, the danger is the mixing between the material of the high-Z shell and the fuel. Because the targets are layered with a high-Z material at the outside and the lighter fuel at the inside, the high-Z material with its large inertial force pushes onto the lighter fuel as the material decelerates. RT instabilities can then lead to a situation where "jets" of high-Z material reach into the fuel. In the worst case this could prevent ignition. But even if ignition does take place, the threat from RT instability effects is not over. Mixing between the fuel and the high-Z material could make the burn much less efficient, thus lowering the gain in the fusion process.

As Duderstadt and Moses (1982) point out, the rule of thumb is that the target should be designed in such a way that ignition occurs before the free-fall line of the interface between the high-Z material and the fuel reaches the hot spot radius.

In contrast to the ablation process, the deceleration RT instability growth cannot be reduced by ablation. The distance of the deceleration is about equal to the radius r_{comp} of the compression. Stabilization occurs mainly as electron conduction establishes a density gradient between the hot and cold material.

Therefore it follows for the constant β in Eq. 6.10 that $\beta \sim 1$. This means

$$\gamma^{decel} = \sqrt{\frac{ka}{1+kL}}. \tag{6.16}$$

For the number of e-foldings n_{max}^{decel} in the deceleration situation it follows that

$$n_{max}^{decel} = \int \gamma_{max}^{decel} dt = \int \sqrt{\frac{ka}{1+kL}} dt. \tag{6.17}$$

Because the gradient is typically 0.1 to 0.2 r_{comp}, assuming constant deceleration over the radius r, $a \sim$ const., and $L \sim 0.2 r_{comp}$ and it follows

$$n_{max}^{decel} \sim \sqrt{\frac{2l}{1+0.2l}}. \tag{6.18}$$

As mentioned above, there are two sources for the perturbations causing RT instabilities — the perturbations at the inside and outside of the shell. The number of e-foldings on the inner surface is reduced compared with

$$n_{max}^{outside} = -k\Delta R. \tag{6.19}$$

As in the acceleration phase, the linearized approach to the deceleration phase breaks down after some time. All that was said about the later stages of RT growth in the acceleration phase — such as saturation and mode coupling — holds in the later stages of the RT instability development in the deceleration phase as well.

To estimate the danger of RT instabilities for the entire ICF process all types of RT instabilities have to be combined — acceleration, deceleration, and feed through. This leads to the total number of e-foldings of

$$n_{max}^{total} = n_{max}^{accel} + n_{max}^{decel} + n_{max}^{outside}, \tag{6.20}$$

where n_{max}^{accel} is different for direct and indirect-drive targets according to Eqs. 6.12 and 6.13.

6.4 Consequences for Target Design

In this section we discuss which kind of perturbation is most dangerous and what this means for the target design.

In the acceleration phase the growth rate is approximately $\gamma = (aK)^{1/2} \sim (al/R_0)$, so that the RT instability increases as

$$R_{per} = R_{per}(t = 0)\exp[\gamma t], \tag{6.21}$$

which means that a shorter wavelength leads to a larger growth rate.

However, as we saw in Section 6.2 very short wavelength disturbances can not be treated with this linear approach any longer. For these short wavelength disturbances the damping is proportional to $\exp(-l/R)$ and they cease to be important

The most dangerous perturbations are the ones where

$$\frac{l}{R_0}\Delta R_0 \sim 1.$$

Avoiding RT instabilities directly influences the target design (i.e., the aspect ratio $R_0/\Delta R_0$), which describes the radius of the target to its thickness. This is easily understood: if the wall is too thin, RT instabilities could in extreme cases completely destroy the shell; too large and the acceleration phase would be longer, giving the RT instabilities a longer time during which to grow.

number of e-foldings to be less than 6 (i.e., $_{max}$ 6), Eq. 6.13 shows that for such target a capsule aspect ratio of $R/\Delta R \leq 25 - 35$ is required with a typical ΔR of $70 - 100$ μm.

If in future it will become possible to produce targets with an even higher surface finish quality, this might allow targets to be manufactured with a higher aspect ratio. As the aspect ratio $R_0/\Delta R_0$ is directly linked to the implosion velocity this would in turn increase the efficiency of the whole implosion process.

6.5   Idealized RT Instabilities vs. ICF Situation

In real ICF implosion experiments RT instability growth is even more complex than described in the previous sections. There are two main differences to the above idealized picture of RT instabilities:

- nonsinusoidal perturbation
- three-dimensional aspects

Beginning with the former, we saw there are many different sources for the initial perturbation of the interfaces between dense and less dense material in the acceleration phase as well as the deceleration phase. The assumption of a sine-shaped perturbation is clearly an idealization — in reality perturbations of very different scales and wavelength will occur simultaneously. The interface typically has a full spectrum with spectral powers at all modes. If such a spectrum of modes is present, one can expect that especially the interaction of the developing structures — that is the saturation phase — will also be influenced.

To illustrate the saturation effect we imagine two perturbation modes. Both modes have nearly equal wavelength, parallel wave vectors and equal amplitude. Because the wavelength is nearly equal, the modes are nearly in phase over a large spatial range. In the regions where they are in phase they add up to a perturbation of the same wavelength but twice the amplitude. The effect of this is that the saturation threshold is reduced by a factor two. Haan (1989) investigated the situation where not only two but many modes are present and defined conditions for the growth rates under these circumstances. These are much more complex than Eq. 6.15.

But even including these saturation effects does not account for the growth of RT instabilities in ICF completely. Realistic models must include continuous gradients in density as well as velocity, time-varying accelerations and energy transport. In addition, one does not have a planar target

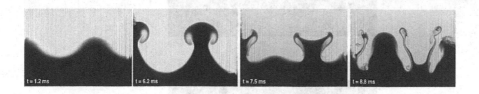

Figure 6.7. Richtmyer-Meshkov instabilities developing in a flow system, in which a light gas and a heavy gas flow from opposite ends of a shock tube (from J. Jacobs, Experimental Fluid Mechanics Group, Univ. Arizona).

but a spherical one, so that three-dimensional calculations are essential. Including all these effects can only be achieved using numerical methods (Dahlberg and Gardner, 1990; Town and Bell, 1991).

6.6 Other Dynamic Instabilities

In the imploding ICF target there are hydrodynamic instabilities other than the RT instability present — namely Richtmyer–Meshkov (Richtmyer, 1960) and the Kelvin–Helmholtz instability (Kelvin, 1910; Dimonte *et al.*, 1993; Hammel, 1994).

Richtmyer-Meshkov instabilities were first investigated in fluid dynamics. These occur whenever a shock wave passes over a nearly planar interface separating fluids of unequal density. Figure 6.7 shows the development of Richtmyer-Meshkov instabilities between flows of a heavy and a light gas. The developing structures are very similar to those of RT instabilities. This is not so surprising as in some way the Richtmyer-Meshkov can be regarded as the limiting case of a RT instability — the gravity acts for an infinitesimally short time on the two fluids and the instabilities develop afterwards even in the absence of the gravitational field.

As before, this fluid description can be directly transferred to the situation in the ICF plasma. In contrast to RT instabilities though, which grow mainly in the two phases of initial acceleration and final deceleration, the influence of the Richtmyer–Meshkov instability is mainly at work in the intermediate stage of nearly constant velocity.

However, in the context of ICF, many more investigations are made into RT instabilities than Richtmyer–Meshkov instabilities because the laser pulses are shaped in such a way that shock heating is minimal. This automatically minimizes Richtmyer–Meshkov instabilities at the ablation surface, too, so that they are regarded as less of a problem than RT instabilities.

Figure 6.8. Kelvin-Helmholtz instabilities. (©The Japanese Society of Fluid Mechanics.)

A hydrodynamic description of Richtmyer–Meshkov instabilities in plasmas analogous to the treatment of RT instabilies in Section 6.1 can be found in Eliezer (2002).

The Kelvin-Helmholtz instability occurs when the two fluids are in motion and encounter a velocity shear, that is a change in the velocity component parallel to the interface between the two fluids. An illustration of the instabilities arising in the case of two fluids is given in Fig. 6.8. In the context of ICF, Kelvin-Helmholtz instability can arise when RT instabilities have already developed. The moving structures (rising bubbles and descending spikes) can induce a shear at the interface because of their motion in turn inducing Kelvin-Helmholtz instabilities. In this case suppressing RT instabilities automatically minimizes Kelvin-Helmholtz instabilities as well.

A second source for the occurance of Kelvin-Helmholtz instabilities in the ICF process are shocks that propagate at an angle to the interface normal. However, the aim of symmetric implosion coincides with minimizing the shear and so the effect of Kelvin-Helmholtz instabilities becomes insignificant Kilkenny et al. (1994). For a more detail description of Kelvin-Helmholtz instabilities we refer to Eliezer (2002).

Chapter 7

Energy Requirements and Gain

We have seen in the previous chapters that almost every process in inertial confinement fusion (ICF) incurs energy losses, which considerably reduces the overall efficiency. In the following we examine what this means for the required the total energy input and the resulting gain.

Because the word *gain* is used in different contexts, we start by providing some definitions. The most important one is that of the target energy gain M given as,

$$M = \frac{E_{fusion}}{E_{d-target}},\qquad(7.1)$$

which is the ratio of the energy E_{fusion} released by the fusion products to the energy $E_{d-target}$ delivered by the driver onto the target. *Scientific breakeven* is reached for

$$M = 1.\qquad(7.2)$$

Achieving scientific breakeven is the main goal of present fusion research programs.

Beyond this point there is also the engineering breakeven, which includes the efficiency of the electrical power needed to run the laser. Engineering breakeven will be discussed in Section 9.2 where ICF power plants are discussed. In this chapter only the processes in the target itself will be considered and not the loss processes in the production of the energy delivered onto the target.

The driver coupling efficiency η_D reflects that only a fraction of the energy the laser delivers onto the target can actually be converted into energy to provide the conditions in the hot-spot area. It is defined as

$$\eta_D = \frac{E_{DT}}{E_{d-target}},\qquad(7.3)$$

core of the target.

In the following we will consider each loss process initially concentrating just on the hot-spot area, gradually broadening the number of factors.

7.1 Power Balance

Here we closely follow the hot spot power balance model derived by Lindl (1989, 1998). A simpler version of this power balance had been suggested earlier by Widner (1979) and Kirkpatrick (1979).

The model by Lindl describes the conditions necessary for ignition to take place in the hot-spot region just before ignition. In this model energy is gained if

$$P_W + P_\alpha + P_n - P_T - P_R > 0. \tag{7.5}$$

This power balance relation expresses that energy is gained by the compressional power P_W, the α-particle P_α, and neutron P_n deposition, but is lost by radiation P_R and electron thermal conduction P_T processes.

Before the void in the target closes in the compression phase, the ion-, electron- and radiation-temperature (T_i, T_e, and T_R, respectively) are the same. Under this assumption, $T_i = T_e = T$, the different terms in Eq. 7.5 can be further approximated:

The specific power P_W of the compression work PdV describes the work done by the pusher per unit volume. Simplifying the model as a piston acting on a gas with uniform pressure, the specific power P_W of the compression work PdV is given by

$$P_W = \frac{P}{V}\frac{dV}{dt} = \frac{P A_{hot} v_{imp}}{V_{hot}},$$

where P is the pressure and v_{imp} the implosion velocity. Because the surface area A_{hot} of the hot spot is $A_{hot} = \pi r_{hot}^2$ and the hot spot volume is $V_{hot} = 4\pi r_{hot}^3/3$, where r_{hot} is the hot-spot radius. This gives

$$P_W = \frac{P(\pi r_{hot}^2)v_{imp}}{(4\pi r_{hot}^3/3)} = \frac{3P v_{imp}}{r_{hot}}$$

Rewriting the latter part of this equation and assuming that the fuel acts as a piston with uniform pressure (i.e., $P \sim \rho R$), this can be reexpressed in terms of the temperature [in keV] and density ρ [in g/cm^3] as

$$P_W = K_1 \frac{\rho T v_{imp}}{r_{hot}} \qquad [\text{W/cm}^3] \tag{7.6}$$

(Frayley   , 1974)

$$P_\alpha = \rho \epsilon_\alpha F_\alpha df_b/dt. \qquad [\text{W/cm}^3]$$

where ϵ_α is the energy of the α-particles per gram of DT

$$\epsilon_\alpha = 0.67 \times 10^{11} [\text{J/g}]$$

F_α the fraction of α-particle deposition and df_b/dt is the rate of change of the burnup fraction, which, according to Eq. 5.43, is

$$df_b/dt = (1 - f_b)\frac{n_0}{2}\langle \sigma v \rangle.$$

Expressed in terms of the mass density this becomes

$$df_b/dt \sim 1.2 \cdot 10^{23} \rho \langle \sigma v \rangle,$$

where n_0 is the total number density of particles and $\langle \sigma v \rangle$ as the Maxwell-averaged reaction cross section. It follows that

$$P_\alpha = K_2 \langle \sigma \rangle \rho^2 F_\alpha \qquad [\text{Wcm}^{-3}] \qquad (7.7)$$

with $K_2 = 2.8 \times 10^{33}$.

The last gain term in Eq. 7.5, the energy deposition by neutrons P_n, is relatively small and therefore can be neglected.

As mentioned above energy is lost from the hot-spot area by electron thermal conduction and radiation. The conduction loss is caused by energy being conducted from the fuel back into the pusher. P_T describes this transfered power divided by the fuel mass. To approximate P_T it is assumed that the conduction losses can be approximated by Spitzer conductivity (Eq. 4.42).

A vital ingredient for determining the conduction power is the knowledge of the temperature profile. This is obtained by assuming a balance between conduction losses and volume heating,

$$\frac{4\pi r^3}{3}(P_W + P_\alpha - P_R) = 4\pi r^2 Q.$$

It follows that $Q = \text{const} \cdot r$. Using the Spitzer-Härm conductivity, it follows

$$Q = -\kappa \nabla T = \frac{-9.4 \cdot 10^{12} S(Z)}{Z \ln \Lambda} T^{5/2} \nabla T \qquad [\text{Wcm}^{-3}] \qquad (7.8)$$

where T_0 is the central temperature of the hot-spot area. This means that

$$T^{5/2} \nabla T(r_{hot}) = \frac{4}{7} \frac{T_0^{7/2}}{r_{hot}}.$$

since $P_T = QA/V$ and $A/Q = 3/r_{hot}$ the electron thermal conduction power is

$$P_T = 3Q/r_{hot}.$$

Using Eq. 7.8 we have

$$
\begin{aligned}
P_T &= \frac{-9.4 \cdot 10^{12} S(Z)}{Z \ln \Lambda} T^{5/2} \nabla T \cdot \frac{3}{r_{hot}} & [\mathrm{Wcm^{-3}}] \\
&= \frac{2.82 \cdot 10^{13} S(Z)}{r_{hot} Z \ln \Lambda} T^{5/2} \nabla T & [\mathrm{Wcm^{-3}}].
\end{aligned}
$$

Taking a the Coulomb logarithm $\ln \Lambda = 2$ and $Z = 1$, this simplies to

$$P_T = q\alpha T^{5/2} \nabla T \sim K_3 \frac{T_0^{7/2}}{r_{hot}^2}, \qquad (7.10)$$

with $K_3 = 8 \times 10^{12}$.

The only term remaining to be determined in Eq. 7.5 is the radiation power P_R. The radiation loss is assumed to be due to radiation leaving the hot-spot area via bremsstrahlung emission processes. This radiation will be absorbed in the surrounding shell, which will again reradiate it. P_R can be approximated by

$$P_R = K_4 Z \rho^2 T^{1/2}, \qquad (7.11)$$

with $K_4 = 3 \times 10^{16}$.

In deriving the different power components in Eq. 7.5 above, several simplifications have been made. For example, the Coulomb logarithm was assumed to be constant, the temperature at the edge of the hot-spot area assumed to be zero, and the temperature profile is more complex than the dependence $T = T_0[1 - (r/r_{hot})^2]^{2/7}$ in Eq. 7.9. For a more realistic quantitative picture, numerical calculations are again neccessary.

Nevertheless, this relatively simple picture allows us to make some predictions central to a successful fusion process. For a constant temperature,

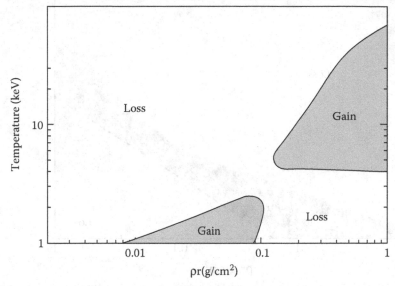

Figure 7.1. Ignition trajectory in the hot-spot concept of the power balance model for an implosion velocity $v = 1 \times 10^7$ cm/s. The boundary between α-particle deposition and compression work PdV is obtained using Eq. 7.13 and the border between conduction and radiation losses is given by $T = 15.8(\rho R)^{2/3}$.

substituting Eqs. 7.6, 7.7, 7.10, and 7.11 in the power balance gives

$$\frac{1}{r^2}\left(K_1\langle\sigma v_{imp}\rangle\rho^2 F r^2 + K_2\rho r T v_{imp}\right) = K_3 T^{7/2} + K_4 Z r^2 \rho^2 T^{1/2}.$$

$$(7.12)$$

In other words, in a region that fulfills the criterion, $P_W + P_\alpha + P_n - P_T - P_R > 0$, we must have

$$\frac{1}{r^2}\left(K_1\langle\sigma v_{imp}\rangle\rho^2 F r^2 + K_2\rho r T v_{imp}\right) - K_3 T^{7/2} - K_4 Z r^2 \rho^2 T^{1/2} > 0,$$

before ignition can occur. This describes that first the energy gain by compression work has to be big enough to obtain ignition and later on the gain by α-particle deposition must sustain and propagate the burn.

Equation 7.12 is an explicit function of $(\rho r,\ T)$. For the boundary $P_W + P_\alpha + P_n - P_T - P_R = 0$, it can be rewritten as a quadratic equation in ρr of the form

$$\left[K_1\langle\sigma v_{imp}\rangle - K_4 Z T^{1/2}\right](\rho r)^2 + (K_2 T v_{imp})\,\rho r - K_3 T^{7/2} = 0.$$

$$(7.13)$$

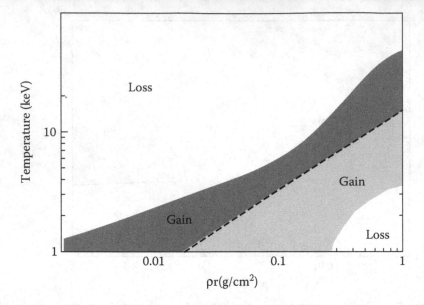

Figure 7.2. Ignition trajectory in the hot-spot concept of the power balance model as in Fig. 7.1 but for an implosion velocity $v = 3 \times 10^7$ cm/s.

Solving this equation for ρr, two solutions exist

$$(\rho r)_{1,2} = \frac{-K_2 T v_{imp} \pm \sqrt{(K_2 T v_{imp})^2 - 4K_3 T^{7/2} \left[K_1 \langle \sigma v_{imp} \rangle - K_4 Z T^{1/2}\right]}}{4 \left[K_1 \langle \sigma v_{imp} \rangle - K_4 Z T^{1/2}\right]},$$

and the area where fusion is possible can be drawn in T and ρr parameter space. Eq. 7.13 is expressed in more practical units as

$$\left[\langle \frac{\sigma v_{imp}}{10^{-17}} F_\alpha \rangle - \frac{T^{1/2}}{2.6}\right] (\rho r)^2 + \left(\frac{T(v_{imp}[10^7 \text{cm/s}])}{35}\right) \rho r - \frac{T^{7/2}}{10^4} = 0.$$

A main result of the solution of the power balance equation is that an in-flight fuel velocity of more than $\sim 2 \times 10^7$ cm/s will be necessary to produce hot-spot ignition. This is demonstrated in Figs. 7.1 and 7.2, where the gain areas are shown for the two implosion velocity of $v_{imp} = 1 \times 10^7$ cm/s and $v_{imp} = 3 \times 10^7$ cm/s, respectively. In Fig. 7.1 it can be seen that the lower implosion velocity, no transition from the low temperature, low ρr gain area to the high ρr gain area exists. In contrast, in the case of a implosion velocity of $v_{imp} = 3 \times 10^7$ cm/s (see Fig. 7.2), uninterrupted trajectories from low ρr gain area to the high ρr gain area are indeed possible. This is the reason why implosion processes are planned in such a way as

electron conduction losses exceed losses by radiation. The two areas are
roughly divided by the line representing the equation

$$T = 15.75(\rho r)^{2/3} \qquad [\text{keV}]$$

in Fig. 7.2. Here the dark gray area indicates the desired gain area for the
compression.

Next we discuss what energy is required to obtain such implosion velocities and the required fusion conditions.

7.2   Energy Requirements

We start with the energy E_{DT} that is contained in the hot-spot region of
radius R. Assuming a spherical volume of DT of density ρ, the number of
atoms is given as $N = \rho V = \rho \times 4\pi/3R^3 = 4\pi(\rho R)^3/\rho^2$. Using the ideal
gas law, E_{DT} is given by

$$E_{DT} = 1.07 \times 10^4 \left(\frac{\rho_l}{\rho}\right)^2 (\rho R)^3,$$

where ρ_l is the density of liquid DT (~ 0.219 g/cm^3), ρR is the areal fuel
density (g/cm^2).

The energy that has to be delivered by the laser onto that target is
obviously much higher, because there are a number of loss processes on the
way to producing the hot spot. What we need is to find the coupling of
the energy into the fuel as defined in Eq. 7.3. The coupling η_D has to take
care of the absorption efficiency η_a, the hydrodynamic efficiency η_H, and
the efficiency of the kinetic to thermal energy transfer η_t. The necessary
energy on the target is therefore given by

$$E_{target} = \eta_D \cdot E_{DT} = \eta_a \cdot \eta_H \cdot \eta_t \cdot E_{DT}.$$

As we saw in the previous chapters that many processes occur simultaneously, so it is impossible to determine η_a, η_H, and η_t analytically. There
is a simple model by Rosen and Lindl (1983) that predicts the required
energy to scale as $v_{imp}^{-10}(P/P_f^3)$.

However, detailed numerical calculations show that this approach is
too simple and predict instead for the *direct drive* scenario the following
dependencies:

$$E^*_{target}[\text{MJ}] - \frac{1}{2}\left(\frac{0.05}{\eta_H}\right)\left(\frac{P}{P_f}\right)^{3/2}\left(\frac{1}{v_{imp}}\right)^5, \qquad (7.14)$$

$$\left(_{\eta}\quad\right)\left(_{f}\right)\ \left(\quad\right)$$

where the temperature is in units of 300 eV.

In these simulations the effects of RT instabilities have not been included, and an ideal hydrodynamic compression was assumed. In reality this will obviously be never achieved, but mixing through instabilities will take place. Including these effects, assuming typical surface finishes of $\sim$ 500–1000 Å, about half the energy is lost through instability effects. In other words, the energy delivered onto the target has to be about a factor of 2 higher,

$$E_{target} \sim 2E_{target}^{*} = \left(\frac{0.05}{\eta_H}\right)\left(\frac{P}{P_f}\right)^{3/2}\left(\frac{1}{v_{imp}}\right)^5,$$

to end up with the same energy in the hot spot area.

For *indirect drive* the situation becomes even more complicated. Here the coupling between hohlraum and target has to be considered, too. Although the conversion of the laser radiation into x-rays is about 70–80%, only part of this radiation actually reaches the capsule. Only 10–15% of the laser input energy to the hohlraum is effectively usable on the capsule itself. This leads roughly to

$$E_d^{hohl} = (7-10) \times E_{target}, \tag{7.16}$$

which, in terms of the required energy input in the capsule, is

$$P_d^{hohl}(\text{MJ}) \sim 350T_r^2(E_d^{hohl})^{2/3}(\text{MJ})\left(\frac{0.1}{\eta_{hohlraum}}\right)^{1/3}$$

$$\sim 350T_r^2(7-10 \times E_{target})^{2/3}(\text{MJ})\left(\frac{0.1}{\eta_{hohlraum}}\right)^{1/3}, \tag{7.17}$$

where again the temperature is in units of 300eV.

It is expected that by optimizing future indirect drive targets, 20–25% of the laser input energy could eventually be transfered into energy usable on the target.

However, the above approximation still underestimates the energy requirements. It ignores the fact that the hohlraum needs to have entrance holes for the laser light to enter, in this way reducing the actual hohlraum inner surface area. The hole areas can be taken into account by approximating the hohlraum coupling efficiency by

$$\eta_{conv}E_{laser} = E_{wa} + E_{cap}^{ab} + E_{hole}, \tag{7.18}$$

$$E_{hole} = 10^{-2} T_h^4 A_{holes} \tau$$
$$E_{cap} = 10^{-2} T_h^4 A_{cap}, \tau,$$

where T_h is the peak hohlraum temperature, A_{holes} the hole area and A_{cap} the capsule area. Here it was assumed that the capsule absorbs all the incident flux.

If one takes into account that some of the energy is reemitted (although this will usually be only a small amount), then the ratio of the capsule to driver energy is given by

$$\frac{E_{capsule}}{E_{driver}} = \eta_{conv} \frac{E_{cap}}{E_{cap} + E_{holes} + E_{wall}}$$

$$= \frac{\eta_{x-ray}}{1 + A_{holes}/A_{cap} + A_{wall}/(2T_r \tau^{0.4} A_{cap})} \quad (7.19)$$

where $\eta \sim 70\text{--}80\%$, T is in units of 10^2 eV, the areas A in mm, and τ in ns.

Because the ratio of the absorbed to the emitted flux is usually small for ICF, the area of the hohlraum walls can be much larger than the capsule itself without losing much of the coupling efficiency. So the relative size of the hohlraum area to the capsule area is primarily determined by the requirement of a uniform capsule illumination. For the targets designed for ignition experiments, the hohlraum area is usually 15–30 times that of the capsule area.

Whenever the capsule reemits some of the incident light, absorption takes longer. It follows that there is more time for losses through the holes and energy absorbed by the walls. Although Eq. 7.19 shows that smaller holes mean less energy loss in the hohlraum, in practice the laser entrance holes should not be too small either. If the laser entrance holes are very small, the laser energy can be refracted and absorbed at the hole edges before entering the hohlraum at all.

In current ignition target designs the ratio of the hole area to the capsule area is 1–2. If the x-ray conversion is approximately 70%, and the target have $A_{holes}/A_{capsule} = 1\text{--}2$ and $A_{wall}/A_{capsule} = 15\text{--}30$, a coupling efficiency of 10–20% exists for such targets.

7.3 Gain

We start with a simple estimate of how the yield Y depends on the driver energy on the target in such a fusion process. The yield can be roughly

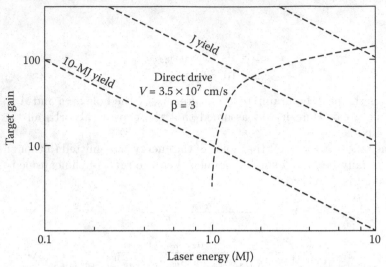

Figure 7.3. Target gain as function of the laser energy for direct-drive targets. Reprinted with permission from Lindl (1995) ©1995, American Institute of Physics.

approximated by

$$Y \sim m_{fuel} E_{DT} f_b,$$

where m_{fuel} is the fuel mass, E_{DT} is the DT fusion energy per gram, and f_b the burn fraction. The burn fraction f_b can approximated by Eq. 5.47. One then obtains for the yield

$$Y \sim m_{fuel} E_{DT} \frac{\rho R}{\rho R + 6(\text{g/cm}^2)}.$$

The fuel mass m_{fuel} in this equation can be approximated by

$$m_{fuel} \sim \frac{\eta_H E_{target}}{1/2 v_{imp}^2},$$

thus leading to

$$Y \sim \frac{2 \eta_H E_{target} E_{DT}}{v_{imp}^2} \frac{\rho R}{\rho R + 6(\text{g/cm}^2)}.$$

For simplicity the hydrodynamic efficiency η_H is assumed to be constant. In Eq. 7.20 the implosion velocity v_{imp} itself is a function of the target

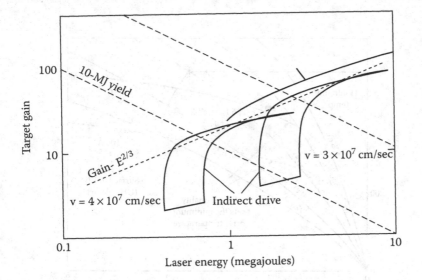

Figure 7.4. Target gain as function of the laser energy for indirect drive targets. For given implosion velosities ($v_{imp} = 3 \times 10^7$ and $v_{imp} = 4 \times 10^7$), the expected gain for optimistic and pessimistic case, assuming a hohlraum coupling efficiency of 10% respectively 15%, is shown (adapted with permission from Lindl (1995) ©1995, American Institute of Physics).

energy again as the relation

$$E_{target} = \frac{1}{2} \left(\frac{0.05}{\eta_H} \right) \left(\frac{P}{P_f} \right)^{3/2} \left(\frac{1}{v_{imp}^2} \right)^5$$

reveals. With this the implosion velocity is given in terms of the target energy by

$$v_{imp} = \left[\frac{1}{2E_{target}} \left(\frac{0.05}{\eta_H} \right) \left(\frac{P}{P_f} \right)^{3/2} \right]^{1/5}.$$

One has to keep in mind that here the implosion velocity v_{imp} is in units of 3×10^7cm/s. It follows that the fuel mass m_{fuel} is related to the energy on the target as

$$m_{fuel} \sim E_{target}^{7/5}.$$

What remains to be determined in Eq. 7.20 is the relation between E_{DT} and the energy on the target. For adiabatic compression of a deuterium-tritium

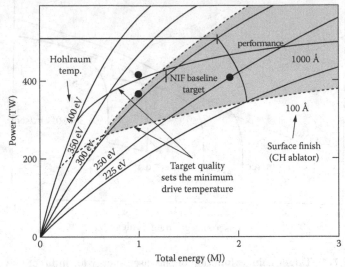

Figure 7.5. The achievable hohlraum temperature is limited by plasma physical issues. On the other hand hydrodynamic instabilities establish the minimum required temperature. The shaded region shows the accessible region in power and energy predicted for indirect drive. Reprinted with permission from Lindl (1995) ©1995, American Institute of Physics.

shell with P/P_f held constant, the specific energy per gram is

$$E_{DT} = \frac{1}{2}v_{imp}^2 = 2\frac{P}{P_f}\rho_f^{2/3}.$$

From this follows that $\rho_f \sim v_{imp}^3$. For a sphere of mass m_{fuel}, it holds

$$(\rho r)_f \sim \left(\rho_f^2 m_{fuel}\right)^{1/3} \sim v_{imp}^3 \cdot E_{target}^{7/5} \sim E_{target}^{1/15}.$$

Combining above equations, it follows for the yield

$$Y \sim E_{target}^{7/5} E_{target}^{1/15} = E_{target}^{22/15}.$$

In deriving this expression we made a few very crude assumptions, more detailed numerical calculations show the following dependencies for a indirect drive target

$$Y[MJ] \sim 375 E_{target}^{5/3} = 375 \eta_{hohl}^{5/3} E_{laser}^{5/3}.$$

As the yield and the gain G are related by $G = E_{laser} \cdot Y$, it follows that

$$G \sim 375\, \eta_{hohl}^{5/3} E_{laser}^{2/3}.$$

can also be read the other way around, indicating the required laser energy for a given gain.

The lines around the areas indicated in grey represent the expected gains for different implosion velocities ($v_{imp} = 3 \times 10^7$ cm s^{-1} and $v_{imp} = 4 \times 10^7$ cm s^{-1}). The gray area represent the uncertainty induced by the deviation from an ideal uniform implosion. The left hand curves relate to an ideal uniform implosion, whereas the right hand lines consider the effect of 500–1000 A deviations in the quality of the target surface. The respective gain curves were obtained under the assumption of a 15% hohlraum coupling efficiency.

Chapter 8

Targets

The first point to realize about targets is that there is no single optimal design for inertial confinement fusion (ICF) research. In fact there are several distinct target designs, each tailored for a different purpose:

- to bring experimental insight into some physical aspects,
- to achieve fusion — like the ones for the National Ignition Facility (NIF) or Laser Megajoule (LMJ),
- targets for a future fusion reactor.

Targets for physics experiments tend to have a completely different design — they might even not be spherical but planar. They are designed to emphasize particular physical characteristics or assist in experimental diagnostics. For example, slab targets are often used to study the interaction of beams with matter in detail: absorption, hot electron generation, transport, hydrodynamic instabilities, and so on. In this chapter we will concentrate on targets designed to achieve fusion and refer to Section 9.4 for a description of the expected alterations necessary for fusion reactor targets. Targets for physics experiments can differ so much from fusion targets, that we will exclude them from the discussion here.

The immediate aim in laser fusion is to achieve ignition and propagating burn with a minimum of incident laser energy. The goal is to design targets that are optimal in this respect. However, the optimum target design depends on many different factors — first of all, on the driver type. The targets for laser driven fusion experiments differ from those driven by heavy-ion or light-ion beams. In this chapter only targets for laser driven fusion will be described, target design in heavy-ion driven fusion is described in Section 10.3.

Even if we consider only lasers as drivers, the indirect and direct approaches still require different targets. The differences will be discussed in detail in Section 8.2. In addition, the optimal target in either of these approaches depends on the individual properties of the laser used. This

in common, which we discuss first.

8.1 Basic Considerations for Target Design

High-gain fusion targets have to fulfill the performance specifications encountered earlier in Chapter 7, which we will demonstrate in the following. Recall that the target design has to suit the driver characteristics — energy, temporal and spatial pulse distribution, number and localization of beams, and focal spot size plus additional parameters specific to the driver type, which for lasers means the wavelength, polarization, and energy spread.

The main tool in designing targets is computer simulation, which models the dynamics of a target in the ICF process. These might be hydrodynamic codes that simulate the whole ICF process or specialized codes that analyze certain aspects relevant for the target design such as:

- Rayleigh–Taylor instabilities,
- transport of suprathermal particles,
- driver energy deposition, or
- energy transport.

To optimize the target design, implosion simulation codes such as LASNEX, HYDRA-3D, ORCHID-2D, and so on are used. For a description of these codes the interested reader is referred to Appendix B.5.

Altogether the target optimization is a highly complex process. However, ICF research has advanced sufficiently in understanding individual processes that some rules of thumb have emerged: the main design parameters are

(i) the amount of deuterium-tritium (DT) — the so-called fuel loading,
(ii) ρR, and
(iii) the shell structure.

The required fuel loading depends on the burn efficiency, which itself depends on ρR. The larger the ρR value, the higher the expected burn fraction of the fuel. This is expressed by Eq. 5.45 in Section 5.9 as

$$f_b = \frac{\rho R}{\rho R + \psi(T_i)},$$

Figure 5.6 shows that a ρR-value of $\gtrsim 1$ g/cm^2 is required for the burn efficiency to be greater than 10% and ρR of 1–3 g/cm^2 corresponds to a compression of the fuel to 300–1600 liquid density.

several mg of DT.

The primary aim of the fusion target design is to compress the fuel to high densities and temperatures with the minimum possible energy input. In other words, the goal is to efficiently push the DT shell to the highest possible velocity in the most efficient way. Therefore providing a long distance over which the shell could be accelerated would be ideal. This would imply making the capsule as large as possible with a very thin layer. However, as we saw in Chapter 7, the problem with that is that any nonuniformity, either by initial surface imperfections or in the heating, is amplified by Rayleigh–Taylor (RT) instabilities. Obviously the longer the acceleration process the more time there is for these instabilities to grow, and in extreme cases the shell might even be destroyed. This puts a limit on the size of the capsule as well as the target manufacturing precision.

Not only does the implosion physics dictate the target design, but equally crucial is the ignition process. In the above described hot-spot scenario, the layer of dense DT compresses a small mass of DT gas in the center. Here again, RT instabilities play an important part. In forming the hot-spot area, the low-density fluid of the DT gas pushes against the high-density fluid originally comprising the solid DT ice layer. A mixing between the two regions sets in and the hot spot is cooled. To avoid the hot spot being cooled below the threshold temperature required for ignition, it is essential that the DT shell is initially very smooth and highly uniform.

At the moment hollow thin-walled spherical capsules of plastic filled with DT gas are the most favored approach to fulfill the above requirements. The DT gas is chilled using cryogenic liquid helium and the DT partly condenses as a thin layer of DT ice on the inside of the shell. A schematic picture of the basic design is shown in Fig. 8.1.

In current designs of direct- and indirect-drive targets, a layer of DT ice is essential. These so-called *cryotargets* operate at a temperature of $\sim$ 18 K and it is important that this is kept relatively constant. If the outer surface of the frozen DT-filled ablator reaches the triple point of 19.79 K, the target may suffer implosion instabilities. For higher temperatures it is even expected that gas bubbles might begin to form inside the capsule (Petzoldt *et al.*, 2002). An alternative to cryogenic targets so-called double shell targets (see Fig. 8.8b) has been considered as a noncryogenic option. However, the designs discussed so far have a much lower performance.

The task for the target designers is to find a configuration that helps to achieve the fusion conditions derived in the previous chapters. These are the required peak temperature and the peak density in the hot spot area, but also the total ρR, the implosion velocity, and others.

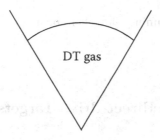

DT gas

Figure 8.1. Schematic capsule design for NIF experiments.

Numerically the target design is optimized the following way: for an indirect-drive target, the parameters hohlraum size, temporal temperature profile, and energy are fixed. The remaining free parameters are ablator thickness and fuel thickness, ablator material, and detailed shape of peak temperature profile. By choosing the ablator material and assuming that the optimal peak temperature profile can be created, this leaves only two free parameters — ablator thickness and fuel thickness. The target designers construct a performance map for this two-dimensional parameter space (see Fig. 8.2) to find the optimal shell thicknesses of the ablator and the fuel. Basically, if the ablator is too thin, it burns through; too thick the

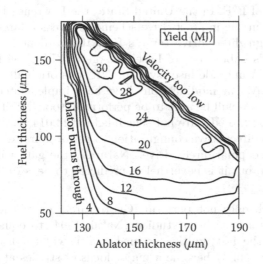

Figure 8.2. Yield of polyimide capsule vs. shell thickness (Haan, 2003). ©2003 by the American Nuclear Society. La Grange Park, Illinois.

target design, the fine-tuning can then be made with two- and three-dimensional codes.

8.2 Direct- and Indirect-drive Targets

As mentioned in Chapter 1, there are two main classes of ICF fusion experiment targets: direct- and indirect-drive targets as illustrated in Figs. 1.7 and 1.10. During the last two decades there has been more research on indirect-drive than direct-drive schemes. This does not neccessarily mean that indirect-drive targets are the better option for a future ICF reactor. Bear in mind that the NIF in the United States as well as LMJ in France are so-called dual-use lasers. That means they are research facilities with the final aim of fusion as an energy source but also used for military research. For these military applications indirect-drive schemes are preferred.

Which scheme is more suitable for an energy-producing ICF reactor is still an open question. We saw in the previous chapters that both direct-drive and indirect-drive targets have advantages and disadvantages and there is still no concensus which scheme would be better in terms of civil energy applications of ICF. In the United States the Lawrence Livermore Laboratory focuses on the indirect-drive scheme, whereas a large portion experiments to design direct-drive targets for NIF are done at the university of Rochester's Laboratory of Laser Energetics (LLE) with the 60-beam, 30-kJ OMEGA-upgrade laser system. NIF was originally designed for indirect-drive only, but modifications have been implemented so that direct-drive experiments will be able to be performed, too. Direct-drive ignition experiments on the NIF are not scheduled until 2014, but there are plans to adapt the indirect-drive configuration to perform first direct-drive experiments — called polar-direct drive. Naturally the gain will be less in this configuration, but it is nevertheless predicted to be ~10 (Skupsky, 2004).

In Japan, which does not perform ICF research for military applications, both schemes have been studied (Nakai, 1994) to equal extent. Recently, however, the fast ignition scheme, which will be discussed in Chapter 11 has increasingly become a major focus of studies at Institute of Laser Engineering (ILE), Japan.

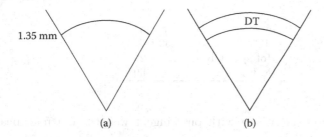

(a)   (b)

Figure 8.3. Schematic capsule design of direct-drive target design for NIF experiments: a) full DT-target; b) foam target.

Direct-drive Targets

For direct-drive targets the main concern is nonuniformity: direct-drive implosions are much more susceptible to radiation nonuniformity triggering hydrodynamic instability than the indirect-drive approach. The main constraints for target design therefore stem from minimizing RT instabilities.

As demonstrated in Section 6.1, Rayleigh–Taylor instabilities determine the upper limit to the aspect ratio — the ratio of the capsule radius R to the shell thickness ΔR. Targets with very thin walls or large dimensions are more vulnerable to instabilities.

This is more specifically expressed by Eq. 6.8, reformulated here as a function of the aspect ratio as (Lindl *et al.*, 1992)

$$\gamma t = \left(\frac{l}{1 + l\Delta R/2R} \right)^{1/2} - \alpha l \left(\frac{\Delta R}{R} \right), \tag{8.1}$$

for a shell accelerated through half the shell radius with constant acceleration and with a density-gradient scale length at the ablation surface optimized to be half the shell thickness. This means that the degree of amplification of perturbations on the surface by RT instabilities is directly determined by the ratio $R/\Delta R$ (Haan, 1989).

Controlling the impact of RT instabilities can be achieved either by minimizing the seeds of these instabilities and reducing the growth rates of the dominant modes. The two main seeds for the instabilities are inhomogeneous radiation and the target surface roughness. The former requires irradiation nonuniformity of 1% rms or less, averaged over a few hundred picoseconds (Skupsky and Craxton, 1999). The techniques used to obtain high irradiation uniformity are spectral dispersion, distributed phase

ρr	1.3 g/cm^2
$\langle T_i \rangle_n$	30 keV
Hot spot CR	29
Peak IFAR	60

plates, polarization smoothing with birefringent wedges, and multibeam overlap (see Section 2.3).

The second constraint leads to target surface requirements. The CH capsules for the NIF experiments are designed in such a way that they can tolerate a surface finish of 800 Angstrom and a mix penetration of 9 μm. Table 8.1 summarizes some of the specifications for direct-drive targets.

However, the right choice of $R/\Delta R$ is not the only option to stabilize the implosion; it is also possible to direct instabilities into a regime in which the density gradients are longer than the instability wavelength. If this can be achieved the instabilities are less damaging for the compression process. In fact the most serious instabilities occur at wavelength comparable to the shell thickness, and these have to be avoided above all else (see Section 6.1).

The growth rate of instabilities can also be reduced by so called adiabat shaping. Here the adiabat of the shell is increased at the ablation surface, resulting in a higher implosion velocity and a reduced instability growth rate γ. The growth rate is kept low in the inner portion of the shell so maintaining the compressibility of the target and maximizing the yield (Goncharov et al., 2003). This shaping of the adiabat can be achieved in two ways: either by a short ($\sim$100 ps) picket pulse creating an unsupported shock in the shell or by relaxing the shell density with a weak prepulse followed by a power shut-off before the main pulse (Andre and Betti, 2004). Figure 8.4 shows how adiabate shaping can reduce the occurance of RT instabilities.

For NIF the direct-drive targets designed by the LLE are cryogenic targets with a spherical DT-ice layer enclosed by a thin plastic shell. As Fig. 8.3 shows there exist two main designs — one with and one without a foam matrix. It is expected from two-dimensional simulations (McKenty et al., 2001) that these targets should lead to a gain $\sim$30 on the NIF. The target shown in Fig. 8.3a is designed for the specifications given in Table 8.1. The design assumes a 1.5-MJ laser pulse consisting of two parts — a fast pulse (4.25 ns at a power of 10 TW) providing a 10 Mbar shock pressure in the DT ice and a second pulse (2.5 ns at 450 TW) timed to

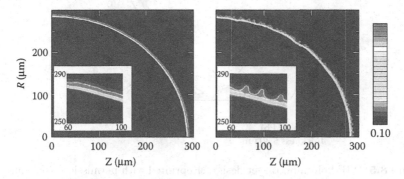

Figure 8.4. Development of Rayleigh-Taylor instabilities in adiabate shaping (McCrory, 2003).

coincide with the shock breakout at the rear surface of the DT ice.

Apart from avoiding RT instabilities, the other major constraint for the target design is to keep the preheat as low as possible. For high gains only a few eV of temperature are allowed in the initial stages of the implosion. Therefore in many target designs the thickness of the fuel damper is determined by the need to keep the preheat low.

The disadvantage of the all-DT target is that, although it is robust, the laser absorption rate is only 60%. By contrast, the DT-wetted foam targets (see Fig. 8.3b) can absorb up to 90% of the laser light. The reason for the increased energy coupling in this type of target is the carbon in the foam, higher Z material, which makes the plasma more collisional. As we saw in Section 4.2 this directly increases the absorption. These factors could push the gain to about ~80, because the higher absorbed energy allows more massive targets to be driven with the same incident laser energy.

These recent advances in direct-drive target design make it more likely that direct-drive ICF could eventually be used for higher gain implosions in an ICF reactor.

Indirect-drive Targets

The basic design of an indirect-drive hohlraum target for NIF experiments is shown in Fig. 8.6. It consists of a cylindrical hohlraum of about 10 mm length, 5–6 mm diameter, and 30 μm gold walls. As noted in Section 4.5 the hohlraum size is not such a critical value, so other shapes and materials are also considered. The important parameter here is the wall/hole surface area ratio, which in current designs is approximately 2:1. The hohlraum is filled with a He or H/He gas enclosed in a 1 μm CH or polyimide windows

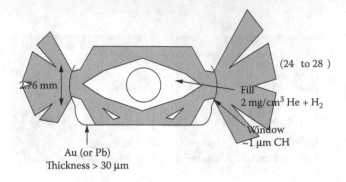

to cover the laser entrance holes. Table 8.2 summarizes the conditions for which NIF targets are designed.

The actual capsule consists of CH shell (see Fig. 8.6a) and its basic design is very similar to that of a direct-drive target: the ablator is a layer of doped CH material with a layer of cryogenic DT and the capsule is filled with DT gas. From Nova experiments it is known that the CH shell contains approximately 5% oxygen as an unwanted fabrication byproduct. The CH shell is doped with 0.25% of bromine to reduce the preheat and improve the stability at the interface between the ablator and the DT fuel.

The gas density in the capsule can be changed by varying the temper-

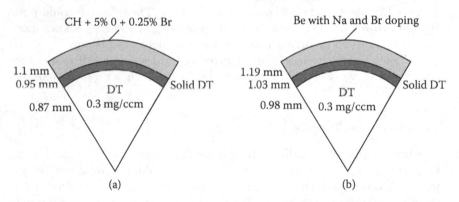

Figure 8.6. NIF capsule designs a) shows a CH-capsule designed for a 300 eV target, which absorbs 1.35 MJ energy; b) shows a beryllium capsule designed for a 250 eV target.

Peak density	1200g/cm^3
Total ρR	1.5 g/cm^2
In-flight aspect ratio	40
Implosion velocity	4.1×10^7 cm/s
Convergence ratio	36
Yield	15 MJ

ature of the cryogenic DT. However, changing the gas content influences the convergence ratio. The yield becomes less for high gas content because the total fuel ρR decreases (see Fig. 8.7). The DT gas in the center will be in thermal equilibrium with the DT ice at a density of ~0.3 mg cm^{-3}.

The target design shown in Fig. 8.6a operates near the expected maximum of the hohlraum temperature of 300 eV. The required laser energy of 1.35 MJ is below the 1.8 MJ the NIF laser is designed to deliver. Likewise the design specifications of LMJ lie above the expected required energy for ignition. This is so that uncertainties in the coupling efficiency and the ignition threshold will not jeopardize the aim of achieving ignition with these facilities. The expected yield of these targets is 15 MJ, this means these targets would have an expected gain of approximately 600. Note that for a fusion reactor a 5–10 times higher energy gain would be required.

There have also been capsules designed which operate at the lower

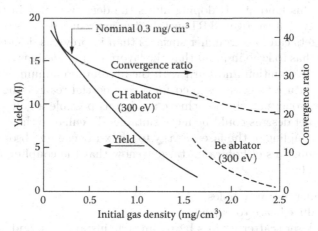

Figure 8.7. Yield as a function of the gas fill for CH and Be ablator targets. Reprinted with permission from Lindl (1995) ©1995, American Institute of Physics.

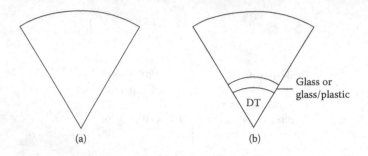

Figure 8.8. Alternative target designs: a) beryllium target with graded copper dopant b) noncryogenic double-shell target.

temperatures. Here the ablator does not consist of CH but beryllium (see Fig. 8.6b for a beryllium target designed to operate at 250 eV). The simulations show that the advantage of these beryllium targets is a somewhat higher achievable ablation pressure. The capsule absorbs more flux at a given temperature. However, beryllium targets are more sensitive to hydrodynamic instabilities and deviations in the pulse. Whether CH or beryllium targets will be used in future largely depends on the achievable fabrication quality of the beryllium targets.

Recently simulations of Be shells with graded copper dopants show very promising results (Haan, 2003). In these targets there is no Cu in the inner and outer edge, but instead increased Cu-doping toward the middle of the shell (see Fig. 8.8a). When the x-ray radiation is absorbed in the capsule, this kind of CU-doping alters the density profile in the target in such a way that the overall RT instabilities are reduced. This in turn means that targets can have rougher surfaces than in previous designs.

One has to keep in mind that the favored target designs are all tailored to achieving ignition and burn with the absolute minimum of incident energy. In future targets designed for a commercial reactor the aim will be a different one; namely, to achieve as high as possible a gain. In this case other target designs could be more suitable. To enhance the target gain an efficient coupling of the laser energy to the x-ray energy absorbed is essential. Simulations of hohlraum targets show that the coupling efficiency be increased by

- smaller entrance holes,
- a reduced case to capsule ratio,
- less laser scattering loss by parametric instabilities, and
- cocktail wall effects in the hohlraum.

The last point is based on the fact that combining appropriate atom-design could increase the coupling efficiency from 8.5% at present to 20% in future.

8.3 Target Fabrication

We saw in the previous section that although the direct- and indirect-drive schemes require two different target designs, the actual capsules are very similar in both approaches — a low-Z ablative material surrounding a dense shell of DT fuel with a 50:50 DT fusion fuel mixture. All target designs share the feature that the fuel is layered, with the DT fuel placed immediately inside the capsule wall.

Consequently, there are several steps required to produce such a target — first a capsule shell has to be made, which functions as an ablator in the fusion experiment, second a layer of DT has to be crafted onto the inside of this shell. For indirect-drive a third step is necessary: manufacture of the hohlraum that surrounds and fixes this capsule at a certain position. Foam targets and doped layers also require additional technical knowhow. In the following the main manufacturing techniques will be described shortly. For a detailed account of recent techniques see Chapter XII in Hammel *et al.* (2004).

Ablator Capsules

In the previous section it was noted that mainly CH plastic (i.e., polymeric composite), and beryllium are favored as shell materials. Most technologically advanced are polymeric composites shells as a result of Nova experiments. The important point in ablator fabrication is to achieve the required roundness and surface smoothness of the capsules and a constant

Table 8.3. Typical specifications for target fabrication.

Specification	value
Deviation from spherical	<0.1 %
Ablator thickness	<1 %
Outer surface smoothness	<200 Å
Inner surface smoothness	<1 μm
Capsule centering in hohlraum	<25 μm
Allowed temperature change of layering	0.5° K
Maximum dislocation in shot (direct)	20 μm
Maximum dislocation in shot (indirect)	200 μm

are dropped down heated towers. The solvents boil and the droplets harden while falling and are collected as microballons at the bottom (Crawley, 1986; Burnam *et al.*, 1987; Cook *et al.*, 1994). A polyvinylacetate (PVA) gas permeation barrier is added and the shell is coated with a polymer ablation layer, which acts as ablator. This capsule fabrication technique was used for most of the targets for Nova experiments. Although a very good surface finish and clarity are achieved, problems remain with the required sphericity and concentricity. Because these defects increase with the size of the capsule, this technique is not suitable for the production of fusion targets for NIF.

The other established technique for capsule fabrication is density-matched microencapsulation mainly developed by the Osaka University Institute of Laser Engineering (Norimatsu *et al.*, 1994), which uses a water in oil emulsion, in which the water forms spherical globules in the polystyrene solution. Adding an additional solution of water and poly vinyl alcohol and heating the whole system drives out the solvent, leaving the polystyrene shells, which harden after several hours (Kubo *et al.*, 2001). With this technique it is possible to achieve larger capsules with very good sphericity and concentricity, but it is less easy to achieve thin walls.

For these reasons the decomposable mandrel technique (Nikroo *et al.*, 2002), will probably be the main capsule fabrication for NIF experiments. Here an inner mandrel is produced by density-matched microencapsulation and coated with a polymer layer. As the capsule is heated, the mandrel material decomposes and permeates out, leaving the shell. Here it is essential that the mandrel is produced with a high-quality surface as the surface finish of the final target depends ion the initial symmetry of the mandrel. Long wavelength surface modulations are reproduced in the final coated shell whereas those of shorter wavelength usually grow in width during coating (McQuillan and Takagi, 2002).

As mentioned before, apart from plastic-based capsules, beryllium and polyimido capsules are also under consideration. The advantage of beryllium targets is they can tolerate a factor 2–3 rougher surface on the outer ablator surface as well as on the inner DT ice fuel surface. However, the fabrication technology is still premature and the deciding factor will be the quality that can be achieved for beryllium shells.

DT Layering

Finally, a layer of DT has to be coated onto the inside of the ablator capsule. The problem here is to ensure that this layer is highly spherical

At the moment solid DT layers are favoured, which means cooling the capsule to low temperatures (Musinski *et al.*, 1980). These targets are called cryogenic by virtue of the fact that the spherical capsule is filled with DT gas cooled using liquid helium. The deuterium and tritium condense as a thin layer of DT ice on the inner surface of the capsule.

Although the solid layer in cryogenic capsules is initially highly nonuniform, fortunately the so-called phenomenon of beta-layering (Hoffer, 1992) smooths out these nonuniformities to a high degree within a reasonable time. In this beta-layering process the small amount of heat created by the radioactive decay of tritium vapourizes the DT ice from thick regions and redeposits it in thinner regions. The manufacturing of these capsules need high precision (Woodworth and Meier, 1997), especially because the plastic shells burst if they warm up too much.

Hohlraum Fabrication

For indirect-driven ICF experiments hohlraum production poses an additional manufacturing issue. The hohlraums are produced by electro-forming gold (Foreman *et al.*, 1994), which involves coating a previously fabricated mandrel with gold and then removing the mandrel chemically. Since the target surface on the inside of the hohlraum has to be very smooth (<100nm rms), special care has to be taken when producing the mandrel. Hohlraum targets also contain the capsule, which is made separately and later butt-joined to the inside.

Chapter 9

Inertial Confinement Fusion Power Plant

This chapter takes a look beyond the inertial confinement fusion (ICF) physics considered before to discuss additional issues that have to be solved in a future ICF power plant. A power station must produce energy at the lowest possible cost. This means the primary objective shifts from demonstrating fusion with the lowest possible energy, to achieving the highest possible gain and plant efficiency.

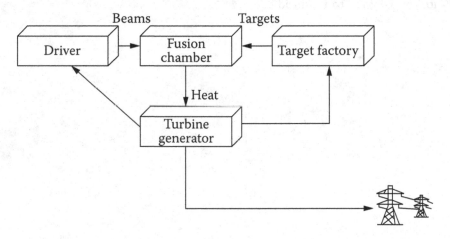

Figure 9.1. Conceptual view of a power plant with separated target factory, driver, fusion chamber, and turbine generator.

in Fig. 9.1.

The different parts of the ICF power plant have the following functions:

- In the **target factory** the targets are prepared and filled with deuterium-tritium (DT). They are made ready to be transported to the reactor and injected into the target chamber.
- The **driver** — whether laser or particle accelerator — converts electrical power into a short, powerful pulse, which then delivers its energy onto the fuel capsule and so drives the fusion process.
- In the **target chamber** the targets position and velocity are tracked and the beams are directed accordingly so that the fusion process can take place. The fusion products are captured in the chamber blanket and their kinetic energy converted into thermal energy. This process is repeated at a high frequency. In addition, tritium is produced in the target chamber.
- In the **turbine generator** the thermal energy in the blanket is converted into electrical power. Part of this electricity is fed back into the reactor to power the driver. The rest can be injected into the power grid.
- In addition, tritium and some other target materials are extracted from the blanket fluid and recycled in the target factory.

The advantage of separating the different parts of the plant is that the target factory and driver are completely decoupled from the fusion chamber, which is susceptible to radiation and shock damage. As we will see in Chapter 10 in the context of the heavy-ion beam driven fusion, an additional advantage of this separation is that a single driver could be used to feed energy to several fusion chambers at the same time. An example of such a IFE power plant — the HYLIFE concept — can be seen in Fig. 9.2.

The construction of a power plant will have to meet the following constringent requirements, in addition to achieving fusion

- High gain targets — i.e., energy output ~50–100 times greater than the driver energy input
- Efficient (10–30%) driver with a high repetition rate (5–10 Hz)
- Low-cost targets with high production rate
- Long lifetime of fusion chamber (~30 years) while maintaining low radioactivity levels

In other words, the driver has to ignite several cheap fusion targets per second with high efficiency for a long time. The ultimate goal is to eventually

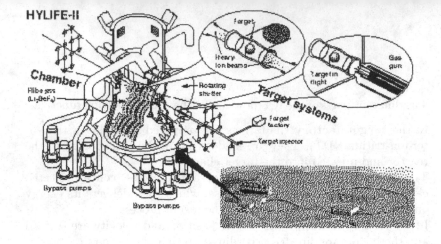

Figure 9.2. HYLIFE concept.

be able to construct a reactor that can provide energy in a relatively clean way without the worry of exhausting resources.

9.2   Plant Efficiency

In a reactor a series of energy conversion processes take place. The reactor efficiency is not only determined by the efficiency of the fusion process (i.e., hydrodynamic efficiency, burn efficiency), but also by how effectively fusion energy can be converted in electricity. The losses and corresponding efficiency definitions are the following:

(i)   The driver efficiency is given by

$$\eta_{driver} = \frac{\text{driver energy input}}{\text{driver electrical input}}$$

and describes the efficiency of converting the input electrical energy into laser light or particle beam energy.

(ii)   The target gain Q is the efficiency with which the driver energy produces thermonuclear energy via the fusion process (see Chapters 4 and 7).

(iii)   When the thermonuclear energy is converted in the blanket of the target chamber, along with the tritium breeding, neutron reactions lead to a modest energy multiplication M.

(iv)   Usually the transformation of fusion energy to electricity will work via

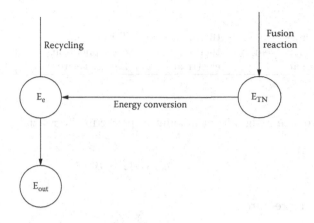

Figure 9.3. Gain cycle for reactor.

a thermal cycle, defined by the gross thermal efficiency of the plant η_{th}.

(v) A certain amount of the electricity produced must be reused to operate the driver. This can be quantified by a plant recycling efficiency η_P.

In Section 7.1 we saw that the scientific breakeven is defined as the point at which the fusion energy production equals the driver energy output. Engineering breakeven is even harder to meet. This is reached if the fusion energy production is high enough to compensate the energy the driver requires, which, for the whole reactor cycle, implies

$$\eta_{driver}QM\eta_{th}\eta_P > 1. \tag{9.1}$$

What values of η_{driver}, Q, M, η_{th}, and η_P seem realistic at present?

Starting with the last one, η_P, the amount of recycled power used to operate the driver obviously should not be too high; otherwise the fuel is just burned to run the driver. For economic reasons not more than 25% of the power produced can be recycled to the driver. Typical thermal cycle have a gross plant thermal efficiency η_{th} between 30 and 40%. The so-called nuclear energy multiplier, the energy increase because of neutron reactions is typically 1.05–1.25, so it will only slightly influence the efficiency of the entire process.

Using these estimates of M, η_{th} and η_P in Eq. 9.1, we obtain

$$\eta_{driver}Q > 10. \tag{9.2}$$

Driver	→ thermonuclear	Target gain	Q	
Thermonuclear	→ thermal	Multiplication	M	1.05–1.25
Thermal	→ electrical	Electr. recirculated	η_P	25%

To produce a significant amount of power in the reactor

$$\eta_{driver}Q \sim 100$$

would be required.

The product $\eta_{driver}Q$ describes the minimum gain necessary for a reactor, independent of the type of driver used. The target gain for the indirect-drive scheme is expected to be $Q \sim 30$ and for direct drive $Q \sim 100$.

As we saw in Section 2.2 for a laser the driver efficiency is realistically about 10%. For the National Ignition Facility (NIF) and Laser Megajoule (LMJ) experiments $\eta_{driver}Q \sim 3$ can be expected; definitely not good enough for a reactor. For a reactor a higher target gain and higher driver efficiency will be mandatory.

One option would be the use of a different type of laser as discussed in Chapter 2, another one to use heavy-ion beams as drivers. As we will see in Chapter 10 the efficiency of heavy ion drivers is expected to be much higher: 30%–40%. Such plants could provide much more efficient reactors.

Any improvement in driver efficiency or target gain would make it possibile to work at smaller driver energies. This would reduce the driver costs as well as require smaller fractions of recycled power. Reduced driver cost and higher efficiency reduce the cost of the energy produced, possibly making it competitive to other methods of energy production in the not too distant future.

Apart from achieving fusion in a cost-effective way, one also has to think about how to convert the energy produced. The energy of the first fusion reactors based on the DT fuel concept will be mainly released in the form of the kinetic energy of the fast 14 MeV neutrons. To capture them a reactor cavity has to surround the pellet and the wall of this reactor chamber must absorb the neutrons and convert their energy into heat, which can then be used to produce electricity.

- provide a good vacuum environment,
- provide an environment inside that the driver can deposit its energy with high precision in space and time,
- make sure that the target can be positioned with high precision,
- recover energy, and
- breed tritium.

The situation is complicated by the fact that all this has to be achieved in a rather hostile environment. A fusion pellet will emit neutrons, x-rays, charged particles, and (for laser drivers) reflect laser light as well.

An essential input for the design of a target chamber is the detailed knowledge of the energy spectrum and yield of the neutrons, photons, and debris. Recent studies show that the amount of x-ray radiation and the energy spectra from the ion debris might differ considerably in the direct- and indirect-drive schemes (ARIES, 2004).

The high-energy neutrons that leave the target and deposit their kinetic energy by collisions will rapidly heat any material in the close vicinity of the target and shatter it. Only at larger distances is the heat deposited more gently.

The x-rays are present because not all energy focused onto the target can be transfered to push the implosion: some of this ends up as x-rays. These are absorbed by the inner surface of the target chamber and vaporize the top layer of the surface, thereby creating debris. In addition there is the debris of the ignited target. The target chamber has to withstand the high velocities of these particles and strong mechanical forces. In a power plant this debris has to be quickly cleared up by condensation before the next target can be ignited. Because of the complex processes of debris deposition, the chamber wall is a key issue in planning an ICF reactor.

Another issue concerns safety aspects, which are directly connected to the choice of the materials used in the target chamber. After many collisions, the neutrons are absorbed by some kind of atom, thereby changing its atomic weight. The newly formed atoms can in some cases be radioactive isotopes, so it is important to choose the materials in the target chamber and the liquid shielding in such a way that radioactive isotopes are avoided, or at least decay in a very short time.

A third point of consideration is the positioning of the targets in the chamber. In current experiments one is able to predict the final position of injected targets to an accuracy of 0.1 mm, a precision thought to be sufficient for ICF experiments (Petzoldt et al., 2002). Because target positioning and beam pointing have to be coordinated, the combined value is

Wet wall chamber	Osiris	Laser	Indirect
	Prometheus	Laser	Direct drive
	KOYO		
Thick liquids	HYLIFE	Heavy ions	
	ZFE	Z-Pinch	
	SENRI		

more relevant, and currently is 0.4 mm (Moir, 1994). Although the positioning does not seem to problematic in single shots, it could be more so if this has to take place in a hostile environment at a rate of several shots per second. We saw in Section 8.1 that the temperature in the target area has to be held relatively constant. This will definitely be more difficult to guarantee if a high repetition rate is required at the same time.

There exist a number of quite detailed concepts (see Table 9.2) for the target chamber design of laser and heavy-ion driven fusion reactors. Fig. 9.6 shows the Sombrero chamber, which is one such fusion chamber concept for direct-drive laser targets. The chamber is made of low-activation carbon-composites. The laser could be a diode pumped solid-state laser or a KrF laser. A xenon gas controls the x-ray and debris damage to the first wall.

In these studies the concepts for laser driven and heavy-ion driven fusion differ to some degree. The final focus and the target chamber have to be sufficiently spatially separated in laser driven fusion, transparent glass shields are required and a large target chamber to reduce the intensity of the x-rays is advantageous. By contrast heavy-ion driven fusion plants can have the final focus magnets closer to the target. This would allow a small target chamber, which would have the advantage of reducing the size of the focus spot. If the focus spot is smaller, the driver can be smaller and is cheaper to build. In addition, a smaller target chamber needs less material and this again reduces the costs. HYLIFE-II (Moir, 1994) is one such study of an IFE power-plant design that uses a heavy-ion driver (see Fig. 9.2). The chamber uses liquid jets of a fluorine, lithium, and beryllium molten salt (Flibe) to protect the fusion chamber from neutrons. This prolongs the lifetime of the components, reduces maintenance costs, and environmental impact.

There are already a number of conceptual designs for the target chamber walls. One can distinguish three primary categories: dry wall chambers, solid wall chambers protected with a liquid film and neutronically thick liquid walls. Table 9.2 shows which wall designs the different IFE reactor studies favor.

Technically complex walls are considered because the fast-flying neutrons basically knock atoms out of their location when they hit the wall. This is more damaging to solid material with a more fixed atomic structure than for liquid matter.

In the wetted-wall concept a renewable liquid provides the shielding of the structural components from the damage by neutrons. Fabrics or thin tubes guide the liquid flow and control the geometry of the liquid. These are also designed to be easily replaceable. Thin liquid films or sheets shield

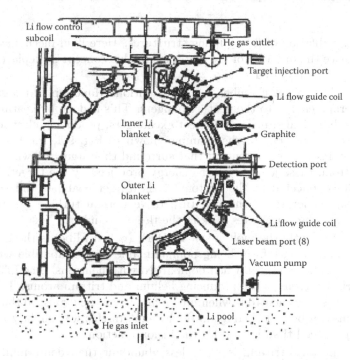

Figure 9.4. Thick wall design of the reactor study SENRI, where the magnetic field guides the thick lithium flow to follow the curved wall (Nakai and Mima, 2004).

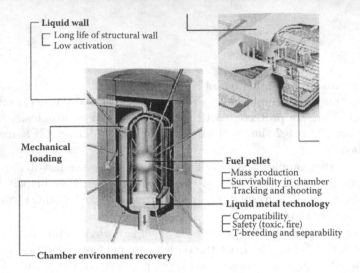

Liquid wall
Long life of structural wall
Low activation

Mechanical
loading

Fuel pellet
Mass production
Survivability in chamber
Tracking and shooting

Liquid metal technology
Compatibility
Safety (toxic, fire)
T-breeding and separability

Chamber environment recovery

Figure 9.5. Liquid wall design in KOYO study (Nakai and Mima, 2004).

x-ray ablation of flow guiding structures. Here fluid mechanical questions replace the question of the material. In this concept no replacement of the blanket is required.

The target chamber has the additional function of transporting the thermal energy to the heat exchangers. This is done by heating a coolant, which is then transfered to heat exchangers, producing electricity via turbine generators. In the example shown in Fig. 9.4, the fusion chamber contains jets of molten salt that surround the target. Flowing from top to bottom, these jets absorb the energy produced by the target. The molten salt is collected at the bottom of the target chamber and transfered to steam generators, which in turn drive normal turbine generators. The coolant then circulates back to the target chamber.

Presently the molten salt Li_2BeF_4 (called Flibe) is the favored liquid coolant. Apart from being nonflammable and compatible with stainless steel, it behaves well when bombarded with neutrons. The Li^6 isotope absorbs neutrons, while producing helium and tritium atoms. The beryllium nucleus loses neutrons when hit by fast fusion neutrons, so that these neutrons can be absorbed again by lithium. In this way slightly more tritium is produced than is used for the fusion reaction.

Because tritium is much less abundant than deuterium, "breeding" tritium in the fusion reactor is regarded as essential. In all existing conceptual studies, tritium is bred from lithium which exists naturally as two isotopes ^6Li (7.4%) and ^7Li(92.6%). Interacting with the neutrons released

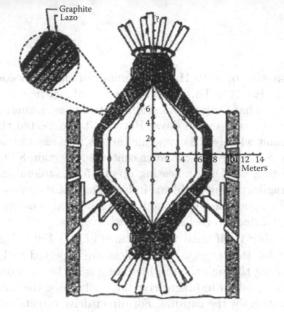

Figure 9.6. Dry wall concept in Sombrero study (Nakai and Mima, 2004).

in the fusion process it can produce tritium in two ways:

$$^7\text{Li} + n(2.5MeV) \longrightarrow\ ^4\text{He} +\ ^3\text{T} + n(slow)$$
$$^6\text{Li} + n(slow) \longrightarrow\ ^4\text{He} +\ ^3\text{T} + 4.8\text{MeV}$$

In designs in which tritium is not bred in the wall, an additional blanket system is necessary to replace the tritium lost in the fusion reaction.

9.4 Target Fabrication for Power Plant

A further critical issue is to manufacture ICF targets economically, especially indirect-drive targets. At the moment it is relatively expensive to produce the targets for ICF experiments, each one costs about $2500 — far too expensive for use in a reactor. The reason for this high costs are that the design of the targets changes constantly and therefore only a few of each design are produced. Development costs are high and the production is labor-intensive.

For an ICF power plant these costs would have to be considerably reduced — estimates range from 20–40 cent per target as the maximum affordable cost (Rickman and Goodin, 2003). This would require cost-cutting but there is optimism that this can be achieved by mass production, having much fewer different designs, changing them less often, and by having

For a repetition rate of 5–10 Hz, the number of targets needed per year for a power plant is $\sim 10^8$. This will involve several logistical problems because, depending on what target fabrication method is used, this involves for example cooling times of several days. Because it is expected that the targets in a power plant will also be cryogenic targets, this means that they must be kept cold (18 K at target chamber center, see Section 8.1) before and during the ICF process. Only a heating up of 0.5 K is tolerable. They are also relatively fragile, posing problems in handling. At the moment the technique of microencapsulation combined with fluid bed coasters seems the most promising fabrication technology to achieve this.

The target chamber itself with temperatures of 500–1500°C poses a hostile environment for the targets. The pellets are expected to become distorted by the heating through the residual hot gas in the chamber. The problem is less pronounced for indirect-drive targets because the hohlraum provides some protection for the capsule. For direct-drive targets reflective metal coatings might help.

9.5 Safety Issues

For a reactor, safety issues will inevitably play a major role. There are several points to consider: the hazard of an accident, spent fuels, activated dust from plasma-facing components, and radioactivity after the shut-down of such a machine or the exchange of certain components.

The chances of a major accident are nearly neglible compared to a fission reactor: melt-down as such cannot occur, because if anything goes wrong the implosion will be unsuccessful and the fusion reactions are simply turned off by themselves. The decay heat from activated material is in the worst case just a few hundred degrees — far too little to melt the target chamber.

What about the hazard posed by tritium? Another advantage of the fusion concept compared to fission is that the amount of radioactive material is much smaller and the half-lives of the fusion products themselves are much shorter — see Section 1.2 — tritium itself has a radioactive half-life of 12.5 years. Furthermore, in the event of an accident with tritium, the timescale of its loss by humans is much shorter. Its biologic half-life is 10–15 days.

Figure 9.7 shows a comparison of the temporal development of the radioactivity in a fission and a magnetic fusion reactor (from a ITER model study) after shutdown. It can be seen that even for a steel construction

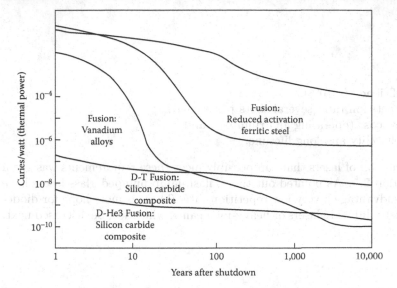

Figure 9.7. Comparison of fission and (magnetic) fusion radioactivity after shutdown ©DOE.

after 20 years the radioactivity is at least two orders of magnitude less. Using more advanced materials, a further reduction of at least two orders of magnitude is expected. In a inertial fusion reactor the radioactivity will be even less because the target chamber is much smaller than a tokamak device.

Because many of the long-term effects on the material in a target chamber are still unknown, a major task of the NIF and LMJ facilities will be to

- obtain data on radioactivity, nuclear heating, and radiation shielding;
- test viability of the wall protection in target chamber;
- measure the dose-rate effects on radiation damage in materials;
- investigate the tritium inventory and tritium burn-up rate; and
- determine the tritium breeding rate.

The results of these investigations will heavily influence the design of a first demonstration reactor, which will be the next technologic step after experiments with the NIF and LMJ laser systems.

Apart from the specifications for research lasers such as NIF and LMJ

- total energy (MJ/pulse),
- wavelength (0.3–0.5 μm),
- intensity (10^{14}–10^{15} W/cm^2 on the target),

- efficiency (>10%),
- repetition rate (several shots per second),
- low cost (operating and running), and
- reliability and long lifetime.

A discussion of lasers that can possibly meet these requirements was given in Section 2.5. As pointed out there, flash-lamp pumped glass lasers have the disadvantage a very low repetition rate. There is more hope for diode-pumped solid state lasers or heavy-ion beams, which will be discussed next.

Chapter 10

Heavy-ion Driven Fusion

While conventional laser-driven fusion facilities (such as National Ignition Facility (NIF) and Laser Megajoule (LMJ)) are essential research tools for studying the possibility of achieving fusion, they are suitable as a basis for an actual power plant. The problems which presently used Nd-glass lasers would face as drivers for a reactor can be summarized by:

- Their efficiency is far too low, typically 6–10%. Even optimzed Nd-glass technology would only achieve 15%.
- A repetition rate of several shots per second would be required, which is presently impossible to achieve with Nd-glass lasers at MJ level — one expects at best one shot every 8 h for NIF.

In Chapter 2 alternatives to Nd-glass lasers with higher efficiencies were discussed. However, there is also the option of using a completely different type of driver — the prime candidates here are heavy-ion beams. Their repetition rate is high and so is the efficiency — for example, an induction accelerator has a typical efficiency of 30%. In this chapter we examine the heavy-ion beam concept more closely.

In heavy-ion beam driven fusion, the material is heated by stopping high-velocity ions in a high-Z material, for example lead. In this case the kinetic energy of the ions is deposited in a very small mass of the material. The advantage of ion beams over lasers is their high efficiency, reliability, and repetition rate. A further advantage of ion beams is that the focusing onto the targets can be done by magnets. Ion beams can penetrate matter with relatively little loss in energy (compared with electromagnetic radiation), so the final focus magnet can much more easily shielded fusion byproducts by appropriate design of the target chamber. There are other problems with heavy-ions as driver as we will see, but let us first start as usual with a description of the driver.

182

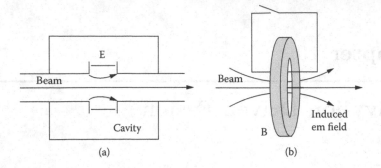

Figure 10.1. Principle of a) radiofrequency and b) induction induced ion acceleration.

10.1 Heavy-ion Drivers

Today there is no complete accelerator designed specifically for the purpose of heavy-ion driven inertial fusion research. Instead, existing accelerators built for high-energy physics experiments — for example, SIS at Gesellschaft für Schwerionenforschung in Germany or TIT/RIKEN/KEK in Japan — are used to perform experiments that are relevant to a heavy-ion driver design or to investigate the interaction of ion beams with plasmas.

The main difference for a heavy-ion driver specially designed for ICF would be to achieve the though requirements for large instantaneous beam power focussed to a small focal spot (~3 mm). This goes far beyond the possibilities of existing conventional accelerators. The basic driver schemes described next are only conceptual, although individual parts of the driver scheme have been tested and there is a proposal to build a integrated beam experiment. Most of the design of such accelerators is presently done using specialized codes.

There are basically two types of accelerators that could be used for heavy-ion fusion

- radiofrequency
- induction accelerator

Radiofrequency (rf) accelerators are mainly favored by Europe and Japan, whereas induction accelerator are favoured by the United States.

In an *rf accelerator* (see Fig. 10.1a) a rapidly oscillating electric field is created by feeding rf-power into a resonant cavity. Short bunches of charged particles are timed to transverse the gap while the electric field points in the right direction to accelerate the particles. The current in an rf accelerator is limited to typically 200 mA — far below inertial confinement fusion (ICF)

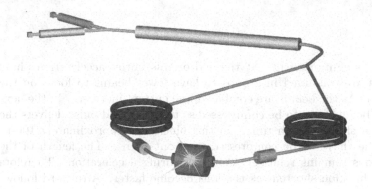

Figure 10.2. Principle of a radiofrequency accelerator for heavy-ion fusion based on the HIDIF study (©Gesellschft für Schwerionenforschung mbH).

driver requirements. Therefore designs of fusion devices usually include stacking beams in storage rings followed by beam bunching in compression rings. A typical layout is shown in Fig. 10.2. Here the charge is accumulated gradually in a series of storage rings.

By contrast, in an *induction accelerator* a changing magnetic field induced by a pulsed voltage creates the electric field that accelerates the ions (see Fig. 10.1b). Although the induction accelerator can handle much larger currents (up to 10 kA) the generated voltage is much lower than in rf accelerators. The schematics of an induction accelerator is shown in Fig. 10.3. Here the beams are accelerated in parallel.

The ion beams are produced the following way: an ion source generates pulses of ions that are initially only singly or doubly positively charged. Currently bismuth, lead, and potassium beams are favored for a heavy-ion

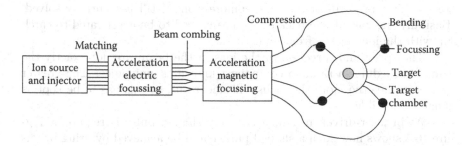

Figure 10.3. Schematic picture of induction accelerator based heavy-ion ignition facility.

beam lines simultaneously. At the end of this initial acceleration phase the beams are combined in groups to have fewer beams to focus on the target area. As the beams are combined, the current increases. In the next phase the beams have to be compressed so that the final pulse delivers the energy in a short enough time. In this phase the approximately 100-ns long bunches have to be compressed to about 10 ns. The length of the cloud of ions remains relatively constant during acceleration. Therefore the pulse duration shortens as the ions become faster. Afterward follows an acceleration with magnetic focusing.

Why does one start with a too long pulse? As we saw earlier, the main difference to conventional accelerators is the high beam intensity. As the ions in the beam are all positively charged, they repel each other, the more so the denser the beam. The result is an unwanted spreading of the beam, which at some point becomes space-charge-dominated. To avoid beam spreading during transport through the accelerator, parts the length of the cylindrical ion cloud is kept relatively long to have a sufficiently low density. In this way repulsion and beam spreading are kept under control. Nevertheless the beam has to be frequently recompressed by electrical fields and magnets.

The space-charge problem is also the reason why in most designs a high number of beams is preferred. If one has a large number of beams, then the current in each single beam can be less, reducing beam spreading and helping to achieve azimuthal symmetry in the annuli when the beam is deposited on the target.

A high total current is necessary when the beam eventually hits the target to have a sufficiently rapid energy deposition. This means that the beam has to be compressed longitudinally by about a factor of 10 just before it hits the target. At this stage the pulse length should be about 10 ns. This problem of the final compression is still not entirely solved. Basically the ions at the back of the beam need to be accelerated to catch up with the ions at the front end.

The speed and direction of the ions in the beam are not exactly the same, but exhibit a random component known as emittance. The greater the emittance the more the focus will smear out, so this has to be kept as small as possible.

As in laser-driven fusion, a carefully shaped pulse is required. Figure 10.4 shows how such a shaped pulse could be achieved by using beams of different duration, current, energies and arrival times (Yu *et al.*, 2005). The numbers in Fig. 10.4 indicate the number of beams of a specific type. In this so-called "robust-point" design the foot pulse of approximately 3

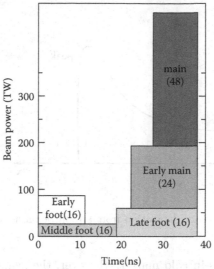

Figure 10.4. Schematic picture of the power profile approximated by constant current beams with different duration (adapted from Sharp *et al.* IFSA, 2003).

GeV is created by three different types of beams and the 4 GeV main pulse by two types of beams. The foot pulse is 15% less than the main pulse as the stopping length of the ions at lower temperatures is shorter (see Section 10.2).

The point design is intended to minimize physics risks. In this design a low-density plasma in the beamline between the final focus magnets and the chamber is used to neutralize the beam charge (Welch *et al.*, 2001). In principle the plasma can increase the beam neutralization to more than 95%. At the moment technical development in ion sources, high-flux injectors, repetitive induction modules and final beam bunching is proceeding quite rapidly (Yu *et al.*, 2003), so we can expect further improvements in these schemes in the near future.

10.2 Ion Beam Energy Deposition

The main differences in energy deposition between ion beams and lasers is that the ions penetrate and deposit their energy well inside the target. Unlike in laser-driven fusion, there is no critical plasma density; instead, the ions are stopped at a well-defined distance. Most of their energy is released near the end of the ion range and very little before. This phenomenon is known as the Bragg peak and is illustrated in Fig. 10.5.

Early work of the stopping of ions in matter concentrated mainly on

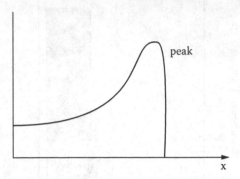

Figure 10.5. The energy deposition from an ion beam as it penetrates matter.

low-intensity beams in cold matter. However, the general concept is very similar to that of high-intensity beams in high temperature material: the ions are slowed down by excitation and ionization processes of the atomic electrons through Coulomb interactions with the ions. This process of stopping of ions in cold matter is described by the Bethe equation as

$$\left(\frac{dE}{dx}\right)_B = \frac{4\pi N_0 Z_{eff}^2 \rho_{st} e^4 Z_{st}}{m_e c^2 \beta^2 A_{st}} \left[\ln \frac{2m_e c^2 \beta^2 \gamma^2}{I_{av}} - \beta^2 - \sum_i \frac{c_i}{Z_{st}} - \frac{\delta}{2}\right],$$
(10.1)

where N_0 is the Avogadro number, $\beta = v/c$, $\gamma = (1-\beta^2)^{-1/2}$. The stopping material is characterized by its density ρ_{st}, its atomic weight A_{st} and atomic number Z_{st}, and the projectile ions by their effective charge Z_{eff}. I_{av} indicates the average ionization potential; $\sum_i c_i/Z_{st}$ characterizes the sum of the effects of shell correction terms. The average ionization potential I_{av} is defined as

$$I_{av} = \frac{1}{Z} \sum_n f_n E_n,$$

where E_n are the possible electronic states and f_n the corresponding dipole oscillator strengths for the stopping material. In practice the average ionization is usually to complex to calculate so experimentally measured values tend to be used instead.

Integrating Eq. 10.1, it follows that the distance r_{st} the ion travels through the matter before it is stopped — the so-called *ion range* — scales

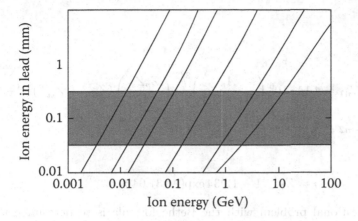

Figure 10.6. Ion energy as a function of stopping distance in lead.

approximately like

$$r_{st} \approx \frac{A}{Z^2}\left(\frac{E}{A}\right)^{1.8} \qquad (10.2)$$

The Bethe formula is only valid as long as $I_{av} < 2m_ec^2\beta^2\gamma^2$; for higher ionization degree it diverges unless atomic shell corrections and polarization effects are also considered. More precise calculations require terms for the contributions from nuclear scatterings of the projectile ion, in which case the stopping power of energetic ions in cold material is more precisely described by

$$\frac{dE}{dx}\Big|_{bound} = min\left(\frac{dE}{dx}\Big|_{Bethe}, \frac{dE}{dx}\Big|_{LSS}\right) + \frac{dE}{dx}\Big|_{nuc},$$

where

$$\frac{dE}{dx}\Big|_{LSS} = C_{LSS}E^{1/2}$$

and the following definitions apply:

$$C_{LSS} = \frac{0.0793(Z_pZ_{st})^{2/3}(1+A)^{3/2}}{(Z_p^{2/3} + Z_{st}^{2/3})^{3/4}A_{st}^{1/2}}\left(\frac{E_L}{1.602\cdot 10^{-9}}\right)^{1/2}\frac{[\text{keV}^{1/2}/\mu\text{m}]}{r_L \cdot 10^4}$$

$$E_L = (1+A)Z_pZ_{st}e^2/Aa$$

$$r_L = (1+A)^2/4\pi ANa^2$$

$$a = 0.468c(Z_p^{2/3} + Z_{st}^{2/3})^{-1/2}\cdot 10^{-8}[\text{cm}]$$

$$A = A_{st}/A_p,$$

$(\rho x)|_{nuc}$ p

$$C_{n1} = 4.14 \cdot 10^6 \left(\frac{A_p}{A_p + A_{st}}\right)^{3/2} \left(\frac{Z_p Z_{st}}{A_{st}}\right) (Z_p^{2/3} + Z_{st}^{2/3})^{3/4}$$

$$C_{n2} = \frac{A_p A_{st}}{A_p + A_{st}} \frac{1}{Z_p Z_{st}} (Z_p^{2/3} + Z_{st}^{2/3})^{-1/2}$$

$$Z_{eff} = Z_{ion} \left[1 - 1.034 \exp(-137.04\beta/Z_{ion}^{0.69})\right].$$

An additional problem with the Bethe formula is to determine the effective charge Z_{eff} of the ions in the beam. For beam ions heavier than protons, the effective Z_{eff} is only known from experimental results that can be approximated according to Brown and Moak (1972) by

$$Z_{eff} = Z \cdot \gamma = Z \left(1 - 1.034 \exp[-(v/v_0)Z^{-0.688}]\right), \qquad (10.3)$$

where $v_O = 2.19 \times 10^8$ cm/s and v the ion velocity in cm/s. According to Eq.10.3 Z_{eff} is essentially a function of the projectile velocity in cold plasmas. The ionization and recombination processes determining Z_{eff} occur on a time scale shorter than the energy loss.

In the ICF context the stopping will mainly occur in high temperature plasmas. For highly ionized targets these equilibrium values of Z_{eff} are often not reached. The energy of the beam ions is more efficiently transfered to free plasma electrons than the bound electrons in cold material. This leads to an increase in the Coulomb logarithm $\ln\Lambda$ and the charge states are higher in ionized stopping material due to reduced electron recombination (Peter and ter Vehn, 1991).

The ion range is shortened and and the Bragg peak even more pronounced than in cold matter. The peak stopping power is enhanced by factors 2 and more from nonequilibrium effects. In Fig. 10.6 the ion range in lead is shown in dependence of their energy. Somewhat counter-intuitively, heavier ions can deposit more energy in a given depth than light ions. Although the beam ions are only singly or doubly charged, as soon as they hit the target, many of the remaining electrons are stripped away. The heavier ions lose more electrons and therefore end up with higher positive charges. These ions can be stopped faster and deposit more energy over a given distance than lighter ions. Thus light ions such as lithium (A = 3) can only deposit 50 MeV within the 0.1 mm of lead, whereas the heavy ions (A = 36–82) can deposit 1–10 GeV. This means the beam intensity can be smaller for heavy-ion beams than light-ion beams.

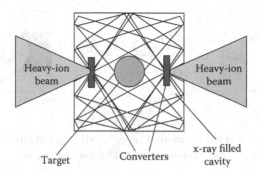

Figure 10.7. Schematic target for heavy-ion driven fusion.

10.3 Target Design for Heavy-ion Drivers

In the heavy-ion fusion concept, not only is the driver different but the energy absorption mechanism described previously also demands an altered target design. Unmitigated, nearly classical Rayleigh–Taylor instabilities (Caruso *et al.*, 1992) are a big problem in a direct-drive targets with heavy-ions as drivers. At present it appears difficult to achieve the required 1% uniformity with ion beams. Nonetheless, direct-drive schemes are still not completely discarded, because new target designs in the laser fusion context provide hope that this requirement could be relaxed. For 2–3% radiation uniformity, heavy-ion driven direct-drive might be possible. However, an indirect-drive scheme seems to be the most likely option.

The indirect-drive concept for heavy-ion drivers differs from that of laser drivers, because the beams do not hit directly the hohlraum walls but deposit their energy in converters (see Fig. 10.7). To directly heat the hohlraum — as in hotraum targets — would require low energy ions and very high peak power. Otherwise the ion energy would be deposited too deep inside the wall and x-ray production would become inefficient. In this case the energy coupling and insulation between absorption and drive stages would become a problem (Piriz and Atzeni, 1994). However, if one uses converters, the energy of the beam ions is transformed into thermal x-ray radiation efficiently. The thermal radiation is confined within the casing and eventually drives the implosion of the fusion capsule. Afterwards everything is expected to proceed as in laser-driven fusion: the typical hot-spot scenario. As before, the capsule consists of a spherical shell of low-Z material and an inner layer of frozen DT fuel.

Because the absorber (i.e. converter) must be heated to 100 eV before it starts emitting x-rays into the hohlraum efficiently, this energy has to

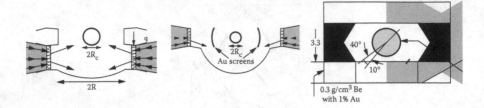

Figure 10.8. Sketch of three different target studies by the HIDIF group (Atzeni *et al.*, 1998). They all assume gold casings and Be converters.

be delivered to the smallest possible volume. This means that both the penetration depth and beam diameter on the target have to be minimized. Target design studies show it is preferable to heat the absorber within 0.1 mm from the surface. Here it is important to use the right kind of ions with the adequate energy to obtain this penetration depth.

Focusing an ion beam to a small spot size is one of the main problems in heavy-ion driven fusion. To focus to the same point after passing through a final focusing device, the ions have to have exactly the same speed and precisely the same direction beforehand. The hotter the beam (i.e., the amount of random motion), the larger the final spot size. This randomness of the motion is called emittance.

Just as in laser-driven fusion, the targets are designed to certain beam specifications. Table 10.3 shows the beam parameters for a two-converter reference target in the HIDIF study (1998). In this study a spot radius of 1.7 mm and a pulse length of 6 ns is the reference point for an ignition driver.

In this study the old idea of telescoping has been revived — here several bunches of different ion species with the same momentum and charge

Table 10.1. Parameters of HIDIF for two-converter reference target.

Parameter	Design Value
Ion energy	10 GeV
Total driver energy	3 MJ
Linac current	400 mA
Storage rings	12
Final pulse length	6 ns
Peak power	750 TW
Focal spot	1.7 mm
Number of final beams	48

Fuel layer thickness	0.05 mm
Fuel mass	0.143 mg
Implosion velocity	4×10^7 cm/s
Peak radiation temperature	250 eV
Nominal energy for ignition	150 kJ

penetrate each other in the final transport and focusing.

As for the laser concept one can describe how efficient the different hohlraum components and accelerator constraints are for a given input energy and one can determine what this means for the energy requirements of the accelerator beam. There are two main processes where energy is lost in the hohlraum target — 1) the conversion of beam energy into x-ray in the converters and 2) the use of this x-ray radiation as driving energy for the implosion. Denoting these two efficiencies transfer η_{tr} and conversion efficiency η_x, respectively, the accelerator energy E_b and the driving energy E_{cap} are connected by

$$E_b = \frac{E_{cap}}{\eta_x \eta_{tr}}. \qquad (10.4)$$

To achieve a high conversion efficiency the volume of converter material heated must be kept small. The heated converter mass m_c can be expressed as

$$m_c \sim N_c \pi r_f^2 R_{ion}, \qquad (10.5)$$

where N_c is the number of converter elements, r_f the size of the focal spot and R_{ion} the ion range in the converter material. In most target designs the number of converters is two, but there are also designs with four or eight converters (see Fig. 10.8).

10.4 Heavy-ion Power Plant

The most expensive part of a heavy-ion fusion power plant would be the ion accelerator itself. This is due to the large size of such an accelerator and the huge amount of iron, copper, and other materials required for its construction. However, a single accelerator could feed a large number of target chambers at the same time by creating 100 or more pulses per second.

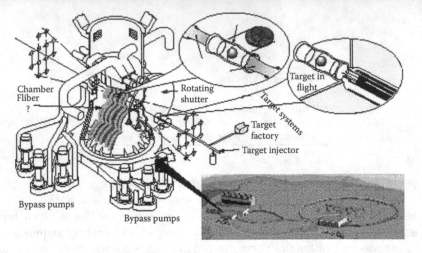

Figure 10.9. HYLIFE design for a future heavy-ion driven fusion power plant ©LLNL and LLBL.

An improvement in gain could be achieved from reducing the case to capsule ratio. This puts higher demands on the brightness and focusing to achieve the necessary small focal spot size.

10.5   Light-ion Drivers

Apart from the heavy-ion beam driven concept, there are also concepts for driving fusion with light-ion beams. Light-ion accelerators use pulsed-power generators and magnetically insulated diodes to produce high currents of light ions such as lithium. Conceptual design studies of inertial fusion energy power plants have been made by the University of Wisconsin and Sandia. The advantage of light-ion drivers is the following: the required ion energy is approximately proportional to the atomic mass. The costs of induction accelerators increase roughly in proportion to the beam energy. Therefore from an economical point of view, low-mass ions would be preferable.

However, light-ion beams have some disadvantages as fusion drivers: the penetration depth depends on the ion mass and energy as well as the absorber material. As we saw earlier heavier ions deposit more energy than light ions per length unit. In other words fewer ions or less intense beams are needed, if the beam ions are of higher Z, which simplifies the beam focusing.

Chapter 11

Fast Ignitor

The fast-ignitor concept was first proposed in 1994 by Tabak, Muro, and Lindl. In their article they suggested igniting the central compressed fuel region by an additional short-pulse high-intensity laser pulse.

In its original form this scheme comprises three phases (see Fig. 11.1):

(i)   the capsule is imploded by a conventional laser to produce a high-density core,

(ii) a hole is drilled through the coronal plasma using a high-intensity 100 ps pulse

(iii) the core is ignited using a 3rd laser pulse with high $I\lambda^2$.

The advantage of the fast-ignitor concept is that compression and ignition are separated, thereby enabling higher gain to obtain for a lower driver energy input, possibly allowing higher tolerances in target fabrication.

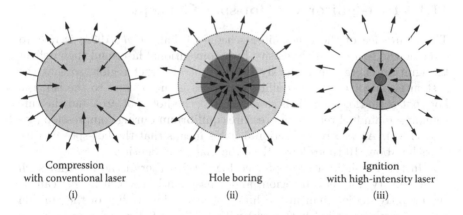

Compression		Ignition
with conventional laser	Hole boring	with high-intensity laser
(i)	(ii)	(iii)

Figure 11.1. Three phases in the fast ignition concept, i) compression, ii) hole drilling, iii) ignition.

194

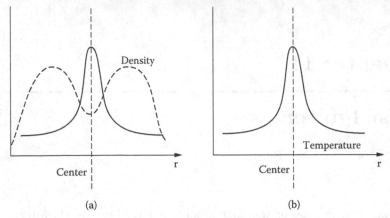

Figure 11.2. Schematic picture of the temperature and density as function of the radius in the a) hot spot concept and b) fast ignitor concept.

Theoretical investigations suggest that in the fast-ignition scheme the target gain could be $Q \sim 200$, whereas the target gain for the indirect-drive scheme is expected to be $Q \sim 30$ and for direct drive $Q \sim 100$. In fast ignition the driver efficiency (if it works well) will be 5%, less than in the other schemes, but because the gain will be higher, this deficiency be more than compensated for.

In recent years there have been a large number of experimental and theoretical investigations to explore the practability of this concept, but it is still too early to say whether fast ignition will work.

11.1 Fast-ignitor vs. Hot-spot Concept

The main idea of the fast-ignitor scheme (Tabak *et al.*, 1994) is first to implode the fuel to a high density by a conventional laser and then deliver the ignition energy very quickly to the central target region, so that the system is out of pressure equilibrium. This is in contrast to the isobaric hot-spot scenario, where it is essential that the hot-spot area and the surrounding main fuel remain in pressure equilibrium during compression (see Fig. 11.2a). As we saw in Section 5.7 this means that the hot-spot density must be about 10 times lower than the main fuel density.

In the fast-ignitor concept (see Fig. 11.2b) there is no need for such a low-density hot central region, and consequently the central ρR can be significantly higher than in the hot-spot case. The radius of the ignition area (sometimes called hot spark) with $(\rho R)_h \sim 0.4$ g/cm^2 can be twice as large as the high-gain fuel radius with $\rho R > 2$ g/cm^2. The advantage of isochoric compression (Tabak *et al.*, 1994) would be that more mass

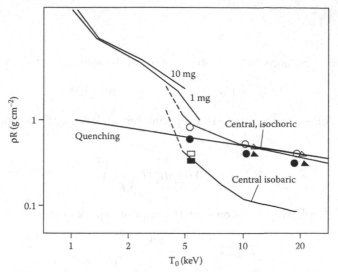

Figure 11.3. Comparison of the ignition thresholds in volume ignition, hot-spot ignition, and fast ignitor ignition according to Nakai and Mima (2004).

could be compressed to much lower density. A larger gain could therefore be achieved because the actual burn mass would be larger or the required laser energy would be smaller for the same gain.

The crucial question for the fast-ignitor concept is whether it is possible to deliver the ignition energy to the pre-imploded plasma. The problem is that the compressed plasma is surrounded by an extended corona. The short-pulse laser somehow has to penetrate this corona and deliver its energy to the overdense regions in the center. The critical density of the created plasma corona lies several hot-spot diameters from the central dense core region. Coupling laser light to supra-thermal electrons at this critical density would result in a very poor efficiency. It is hoped that making use of relativistic effects and the ponderomotive pressure, a way could be found to bring the intense light much closer to the compressed core. There are two main suggestions how to achieve this: hole boring or laser cone guiding (see Section 11.2) .

Assuming the energy does reach the overdense region near the center, then for the moment that supra-thermal electron energy must be converted into thermal electrons, from there to the ions and eventually to the kinetic energy of the fuel.

There are two sources required to provide energy input — the conventional laser system and the high-intensity, short-pulse laser system. Consider now the energy requirements for the compression:

the ignition conditions in the isobaric case can be approximated by

$$(\rho R)^3 T = 1.0 \qquad\qquad [(g/cm^2)^3 keV] \qquad\qquad (11.1)$$

when the ignition temperature is approximately 10 keV, with the hot-spark energy as

$$E_{spark} = 10.6 \, (\rho R)^3 T \left(\frac{\rho}{\rho_s}\right)^2$$

in $[(g/cm^2)^3$ keV GJ]. It follows for the required spark energy

$$E_{spark} \sim \frac{40kJ}{(\rho/100 \text{ g/cm}^3)^2}. \qquad\qquad (11.2)$$

If one wants to compare the two models, the total energies in MJ in the compressed system is given

$$E_{hot\ spot} = \frac{5.8 \times 10^6 T_{hotspot}^3}{\alpha^2 \rho_M^{10/3}} + 0.35 \alpha M \rho_M^{2/3}$$

$$E_{fast\ ignitor} = \frac{0.031T}{\rho^2} + 0.35 \alpha M \rho^{2/3}.$$

Figure 11.4 compares the gain, in-flight-aspect ratio and convergence ratio for the two concepts, and Table 11.1 shows a comparison of the optimum quantities according to Tabak *et al.* (1994).

The short-pulse laser energy input depends on the heating depth, the diameter of the heated area, and the absorption efficiency. The requirements on the pulse duration of the ignition laser are determined by the

Table 11.1. Comparison of the optimal quantities in hot-spot and fast ignitor concept as a function of the internal energy of the initial system ηE in MeV according to Tabak *et al.* (1994).

Property	Hot-spot model	Fast ignitor model
Gain	$1.5 \times 10^3 \eta(\eta E)^{0.3}$	$3 \times 10^4 \eta(\eta E)^{0.4}$
$R_{hotspot}(\mu m)$	$190(\eta E)^{0.5}$	$120(\eta E)^{0.5}$
$R_{mainfuel}(\mu m)$	$280(\eta E)^{0.5}$	$1200(\eta E)^{0.6}$
$\rho_{mainfuel}(g \text{ cm}^{-3})$	$358(\eta E)^{-0.3}$	$33(\eta E)^{-0.5}$

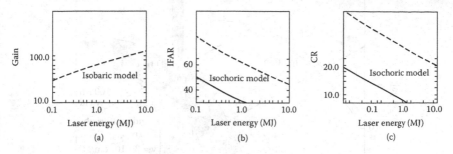

Figure 11.4. Comparison of a) gain, b) in-flight-aspect-ratio and c) convergence ratio plotted as a function of the laser energy in hot-spot and fast ignitor concept (Tabak *et al.*, 1994).

energy flow and disassembly time of the ignition region. The pulse length at peak intensity should be between the electron–ion coupling time τ_{ei} and the disassembly time of the fuel τ_D. Following Tabak *et al.* (1994) and assuming a DT density of 300 g/cm³, 5 keV temperature, $\rho R = 0.4$ g/cm², and a sound speed of 1 μm/ps, the condition for the pulse duration is given by

$$\tau_{ei}(\sim 10^{12}\text{ s}) \leq \tau_{pulse} \leq \tau_D(\sim 10\text{ ps}). \tag{11.3}$$

Assuming that the final fuel temperature is 10 keV, the required energy in the fuel for the above case is 3 kJ and the intensity 8×10^{19} W/cm². Any coupling inefficiencies have been neglected in this estimate, the most serious of which is how efficiently the energy can be coupled by the ignition laser into the fuel.

It is believed that this whole picture (especially Table 11.1 and Fig. 11.4) has been too optimistic. However, for a laser energy of about 1 MJ the isochoric gain is still expected to be nearly three times greater than the isobaric one (Atzeni, 1995).

At the moment the fast-ignitor concept is still very much at the proof of principle stage. At Institute of Laser Engineering, Osaka, the Rutherford Appleton Laboratory and LULI lasers of a few 100 TW can routinely deliver up to 100 J in 1 ps. These lasers will soon be upgraded to the petawatt level [1] started delivering 0.5–1 kJ/ps and will be capable of generating hot e⁻ currents of several MAmps.

Many variations on the original Fast Ignitor concept have already been put forward, including a hybrid of heavy-ion driven compression with fast

---

[1] The VULCAN laser at Rutherford Appleton Laboratory achieved petawatt power on 5 October 2004.

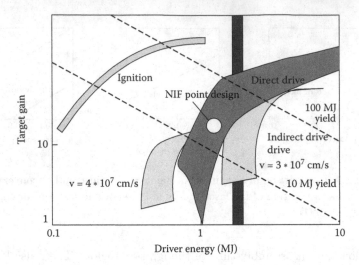

Figure 11.5. Comparison in target gain in direct drive, indirect drive, and fast ignition.

ignition (Caruso and Pais, 1996; Atzeni *et al.*, 1997) as well a fast ignition using laser-accelerated protons (Roth *et al.*, 2001).

11.2   Hole Boring or Laser Cone Guiding?

One of the main problems in the fast-ignitor scheme is the transport of energy to the high-density regions of the plasma. There are two methods under consideration — hole boring and cone guiding.

When the hole-boring laser pulse hits the target, because of the long-scale coronal plasma the coupling efficiency depends strongly on the focus position of the heating laser (Tanaka *et al.*, 2000). Experiments by Kitagawa (2002) showed that the laser pulse penetrated the overdense region when the focal position was near the critical density. The result was an enhanced neutron yield, but this was dominated by the high-energy ions and not the thermal ones and it was not clear whether the enhanced neutron yield came from fusion reactions near the critical surface or from the core plasma as intended. However, it is anticipated that for pettwatt laser energies of 1 kJ and above, the laser pulses might penetrate to higher density regions. The reason is that the strong plasma heating might cause the nonlinear scattering to saturate. In this case the thermal neutrons might increase as well and so might the core plasma heating.

In recent years a target design of a capsule with a guiding cone (see Fig. 11.6) has become increasingly popular for fast ignitor studies, for the

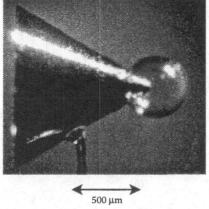

500 μm

Figure 11.6. Cone target Nakai and Mima (2004).

simple reason that hole-boring issues are sidestepped. The cone is made of a high-Z material such as gold, because then the cone wall remains intact when the plasma develops. The fuel shell is imploded to produce a compressed core plasma near the tip of the cone. For a reactor-sized cone target, simulations predict a density of 1000 times solid density and a ρr higher than 2 g/cm². When the high density is reached the heating pulse will be injected at the time of maximum compression.

In cone shell targets the laser energy is mainly deposited at the top of the cone. The hot-spark radius could be influenced by the choice of the cone top radius and the distance between the cone top and the core plasma. In recent experiments about 25% of the short-pulse laser energy were transported to the core plasma in such cone targets (Kodama *et al.*, 2001).

11.3 Off-center Ignition

In most simulations it is assumed that the ignition occurs right in the center of the high-density area. Because delivering the energy far inside the dense plasma seems to be difficult, there have been investigations (Piriz and Sanchez, 1998; Mahdy *et al.*, 1999) considering the much more likely case of an so-called off-center ignition, where the ignition occurs not right in the geometric center but still in the high-density area (see Fig. 11.7). As Fig. 11.2 indicates they find that at least in two-dimensional simulations, the ignition conditions (represented by Eq. 11.1) are approximately the same for central as for off-center ignition.

However, this does not mean that the required spark energy is the same. The reason is that the hot spark is produced at the edge of the fuel by relativistic electron heating. As Deutsch *et al.* (1996) showed the stopping

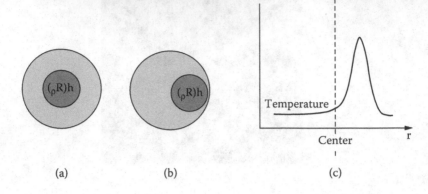

Figure 11.7. Schematic picture of a) central spark and b) off-center spark ignitionin the fast-ignitor scenario; c) shows the temperature and density distribution in an off-center spark ignition situation.

range of the intense electron beam is shortened under these conditions. Because of this stopping range shortening and the heating geometry the required spark energy scales according to Atzeni (1999) as

$$E_{spark} \sim \frac{140 \text{ kJ}}{(\rho/100 \text{ g/cm}^3)^{1.85}}. \tag{11.4}$$

The required energy can be therefore more than three times higher than for central ignition.

11.4   Status and Future

From the current understanding of the heating processes, ignition will be achieved with a pulse energy less than 50 kJ in 10 ps for an imploded plasma ρR higher than 1.0 g/cm^2 (Nakai and Mima, 2004).

There are a number of short-pulse lasers under construction to study the fast-ignitor concept in more detail. Among them the OMEGA EP (extended performance) which will add two short-pulse beams of 2–3 PW and 2.6 kJ to the existing OMEGA laser system. The focussed intensities are expected to be up to 6×10^{20} W/cm^2. and is scheduled to operate in 2008.

Chapter 12

ABCs of ICF

Ablator   The ablator is a shell of material surrounding the outside of the fuel capsule. It is quickly heated by the driver beams, evaporates outward and because of momentum conservation, accelerates the fuel toward the capsule center.

Absorption   Most of the laser light is absorbed at or below the critical surface. In laser-driven fusion the main absorption process is inverse bremsstrahlung absorption, which involves collisions between plasma electrons and ions.

Acceleration phase   Part of the ICF process, in which the fuel moves inward with increasing velocity as the driver deposits its energy.

Aspect ratio   Ratio of the capsule radius to the shell thickness. The **in-flight aspect ratio** is the aspect ratio during the compression phase.

Attenuation factor   The attenuation factor describes how many hot electrons reach the fuel as a function of the thickness of the ablator and the mean range of the hot electrons.

Beam shaping   For both laser and heavy-ion-driven fusion, the beam has to have a temporal and spatial shape to ensure that the compression can follow an adiabat as closely as possible (see **pulse shaping**).

Bragg peak   As ions penetrate into material most of their energy is released near the end of the ion range. This phenomenon is called the Bragg peak.

Breakeven   Scientific breakeven is defined as the point at which

backscattered and the laser absorption decreases.

Burn fraction   In any given configuration it will never be possible to burn the entire fuel. The burn fraction f_b gives the ratio of the burned fuel to the total fuel.

Collisional damping   Under certain conditions, collisions in a high-Z plasma can damp excited waves sufficiently that instabilities (such as Raman) can be stabilized.

Compression   In the hot-spot concept the aim is to achieve an adiabatic compression because this is most energy efficient. By contrast, the fast ignitor concept uses isochoric compression.

Confinement time   The fuel must be confined for a sufficiently long time high enough density, so that a sufficient number of fusion reactions take place. For inertial confinement this time is in the picosecond range, whereas for magnetic confinement, several hours are envisaged for a fusion reactor.

Convergence ratio   Ratio of the initial to final capsule radius after the implosion.

Coulomb logarithm   The cross section for electron-ion scatterings integrated over all velocities and angles contains a logarithmic term that reflects the limits of the integration — namely the Debye length and the distance of closest approach.

Coulomb repulsion   That equal charges repel each other. In the fusion process the Coulomb repulsion has to be overcome to fuse two nuclei.

Critical density   The plasma hinders the laser beam from penetrating regions with densities higher than the critical density. In laser-driven ICF experiments the critical density surface is located at some distance from the solid target surface, so the laser energy is typically deposited several microns in front of the target surface.

Debye length   Because of the attractive and repulsive forces in plasmas, an ion will be surrounded by electrons and vice versa, so that on a large scale the plasma will be quasi-neutral. The Debye length λ_D

netic wave decays into an ion and an electron wave.

Deceleration phase   The deceleration phase is the part of the implosion in which the fuel approaches the center, is slowed down, and reaches its final stages of compression.

Degeneracy parameter   Parameter that describes the degree to which quantum mechanical effects begin to dominate the behavior of the electron gas at high densities and low temperatures, so that Fermi-Dirac rather than Maxwell-Boltzmann statistics apply.

Deuterium   Deuterium is a naturally occurring isotope of hydrogen. In constitutes one proton and one neutron. Deuterium and tritium are the preferred fuel combination for current fusion devices because of their large fusion cross section.

Dilution factor   Geometric factor that represents the considerable amount of hot electrons that do not hit directly the core but reach it after many scattering events.

Direct drive   The power of the driver is directly deposited on the surface of the fuel capsule and drives the implosion.

Disassembly time   Time the target needs to fly apart so that no further fuel can be burned.

Driver   Machine that produces the required laser or ion beams which deposit their energy on the target.

DT reaction   The fusion reaction of deuterium and tritium represents the easiest approach to fusion becuase of a relatively large cross-section and very high mass defect. When these two nuclei fuse, an intermediate nucleus consisting of two protons and three neutrons is formed, which splits immediately into a neutron of 14.1 MeV energy and an α-particle of 3.5 MeV.

Electron conduction   In the burn phase electrons diffuse into the surrounding colder plasma and reduce the temperature in the hot spot.

Emittance   The ions in a beam never have exactly the same speed

by a conventional laser to produce a high-density core; there, the core is ignited using a short-pulse laser with high intensity.

Final focus   The last focusing element of a driver beam that focuses the beam diameter to the required spot size on the fusion target.

Flux inhibition parameter   The actual measured thermal flux in plasma coronae is typically an order of magnitude smaller than the values obtained from theoretical models. An artifical inhibition parameter is used to deal with this inconsistency in hydrodynamical models.

Fusion reaction   The **DT reaction** is the easiest fusion reaction to achieve. Other reactions that occur naturally in the sun and stars would require a larger energy input.

Gain   The energy gain in ICF is defined as the ratio between the fusion energy produced and the total energy put into the driver beams.

Hohlraum   The hohlraum is an enclosure around the ICF capsule used in the indirect-drive approach. Here the driver beams do not strike the capsule but the inner walls of the hohlraum.

Hohlraum coupling efficiency   In indirect drive the laser irradiates the hohlraum wall and the laser light is converted into x-ray radiation. The hohlraum coupling efficiency is the ratio of the x-ray radiation energy to the energy of the laser light.

Hole boring/drilling   In the fast ignitor concept a high-power laser has to drill a hole through the plasma to at least the critical density of the target.

Hot spot   The inner part of the fuel is compressed into a higher-temperature adiabat than the outer part of the fuel. Both parts are compressed to high densities, but the hotter inner part is slightly less dense than the outer part.

Hydrodynamic efficiency   The hydrodynamic efficiency takes into account that the absorbed laser energy goes into ablation as well as the acceleration of the target. Therefore only a part of the total energy PdV applied to the implosion can be used for the compression.

confines the fuel sufficiently long that energy surplus is possible.

IFE (Inertial fusion energy)   Program with the aim of constructing an energy-producing reactor.

Implosion velocity   A high implosion velocity is essential for an efficient ICF process. Together with ρR it determines the achievable gain. In current ICF fusion experiments velocities of $\sim 3 \times 10^7$ cm/s are typical implosion velocities.

Indirect drive   The driver does not deposit its energy on the fuel capsule but is first converted to x-rays in a casing surrounding the capsule known as 'hohlraum'.

Induction accelerator   Linear accelerator which accelerates particles using rapidly changing magnetic fields.

In-flight aspect ratio   The ratio of the shell radius to the shell thickness during all stages of the compression is called the in-flight aspect ratio.

Injector (beam)   The ion source and the first stage of a particle accelerator.

Injector (target)   In a reactor many shots per second will have to be performed. Here it is necessary to inject a new target into the target chamber after each shot.

Instability   Any process where a small perturbation grows (initially) exponentially. See also *Parametric instabilities* and *Rayleigh-Taylor instabilities*.

Inverse bremsstrahlung absorption   If an electron oscillating in a laser field is scattered in the field of an ion and it will absorb a number of photons.

Ion range   While penetrating an absorber, ion beams deposit nearly all their energy at a well defined depths — the ion range.

Kelvin-Helmholtz instability   A Kelvin-Helmholtz instability oc-

Laser principle   Energy is pumped into the laser medium, the atoms inside become excited. As they decay they emit photons, which might hit an other excited atom. This atom then emits a photon exactly in phase with the first one. Such repeat processes lead to an amplification of light, where all photons travel in the same direction in phase.

Lawson criterion   If $3nk_BT < \frac{n^2}{4}v\sigma\tau Q$, the fusion reactions release more energy than is required to produce the plasma of such temperature and density. This relation is called the Lawson criterion.

Liquid shielding   High-energy neutrons leave the target and deposit their kinetic energy by collisions. Liquid shielding of the chamber walls reduces the structural damage of the walls.

LMJ (Laser Megajoule)   Proof-of-principle experiment using a powerful Nd-Laser to ignite a fusion target (located in France).

Magnetic fusion   Controlled thermonuclear fusion approach where magnetic field confine the plasma.

Mode coupling   In the later stages of the Rayleigh-Taylor instability growth the "bubble-and-spike"structures no longer grow isolated from each other, but start to influence each others' growth. This effect is called mode coupling.

Neutron deposition   The energy deposition by neutrons is relatively small and can usually be neglected in energy considerations. However, it contributes to the damage of the chamber walls.

NIF (National Ignition Facility)   Proof-of-principle experiment using a powerful Nd-Laser to ignite a fusion target (located at Lawrence Livermore Laboratory, USA)

Parametric instabilities   The resonant decay of an incident wave into two new waves is called parametric instability.

Plant efficiency   For a power reactor the efficiency is determined by the way the fusion energy can be converted in electricity. Usually this

Ensemble of chemically unbound ionized particles.

Plasma frequency   If the charge neutrality of a plasma is violated, the electrons respond in order to restore it. This can result in an oscillation with a frequency depending on the square root of the plasma density — the so-called plasma frequency.

Power balance   The power balance relation for the hot spot expresses the fact that energy is gained by the compressional power, the α-particle, and neutron deposition, but lost by radiation and electron thermal conduction. It can be used to determine gain areas in $(\rho r, T)$-plane.

Prepulse   A low-power pulse before a succession of increasingly intense pulses is used to accelerate the fuel nearly isentropically.

Preheat   Preheat is the premature heating of the fuel by fast (or hot) electrons. Preheat makes compression more difficult.

Pulse shaping   To achieve an energy efficient compression, the laser pulse has to deliver energy onto the target with a particular time-history. In addition there exists spectral and spatial pulse shaping.

Radiation uniformity   Degree to which the capsule can be illuminated uniformly over its entire surface. The degree of radiation uniformity determines how successful the compression phase will be. Nonuniform illumination occurs on both microscopic and macroscopic scales. Macroscopic nonuniformity can be either caused by a too small number of illumination beams, or the existence of a power imbalance between the beams. Microscopic nonuniformities form from spatial fluctuations within a single beam.

Raman scattering   If the laser frequency is larger than twice the plasma frequency, a scattering process called stimulated Raman scattering can occur. This involves the decay of the electromagnetic wave into an other electromagnetic and a electron plasma wave.

Rayleigh-Taylor instabilities   These instabilities can occur when a denser material pushes onto a less dense one. If this metastable state is perturbed, a mixing between the two regions can set in. In ICF Rayleigh-Taylor instabilities can occur in the acceleration and deceleration phase and represent one of the biggest threats to successful compression.

Resonance absorption   Important (but in ICF undesirable) absorption process of laser light near the critical surface of the plasma. Because of the steep density gradient there, electromagnetic waves are resonantly excited transferring energy from the laser light into plasma waves. Because these waves are damped, this energy will eventually be converted mainly into fast electrons.

Richtmyer-Meshkov instability   Richtmyer-Meshkov instabilities occur whenever a shock wave passes over a nearly planar interface separating fluids of unequal density.

Rocket model   In a rocket the steady blow-off of the fuel accelerates the rocket. Similarly, in the ICF process the conduction of heat into the ablation front builds up pressure that drives the ablation of matter from the outside of the capsule. This in turn leads to the acceleration of fuel in the opposite direction — toward the center of the capsule — and drives the implosion of the target.

Saturation   In the latter stages of Rayleigh-Taylor instabilities, the perturbations no longer grow exponentially but linearly. This phase is called saturation.

Self-heating   If the fuel density in the hot-spot region is sufficient, the fusion products are stopped, deposit their energy, and the temperature increases. This is called self-heating, which in turn allows more fusion reactions to take place.

Shell structure   The deuterium-tritium-containing capsule usually consists of several shells of different material. This is called the shell structure.

Shock wave   In a plasma disturbances propagate faster in high-density regions than in low-density regions. When the fast propagating disturbance travels into a lower density region, the perturbation profile will steepen and eventually develop into a sharp wave front: a shock wave.

Space-charge-dominated beams   Ion beam in which the effective electrical repulsion force of the ions is stronger than the pressure associated with the internal temperature of the beam.

Target   ICF fusion targets contain the fuel that is compressed and eventually burns. Usually they consist of a spherical capsule that contains DT and has a outer layer of higher-Z material as ablator. Indirect drive-targets are additionally suspended within an enclosure — the so-called hohlraum.

Target chamber   The target chamber encloses the last stage of an ICF complex — that is, the final focusing system and the actual area around the target. It has several functions: it encloses the target and creates a vacuum around the target; protects the surrounding from damage by the produced neutrons, photons, and debris; and extracts the gained fusion energy.

Thermal conduction   The energy that is absorbed at the critical surface is transported towards the solid target either by radiation or electron thermal conduction. The heat conduction process is dominated by the much lighter and faster electrons.

Tokamak   Large, torus-shaped fusion device surrounded by electrical coils producing magnetic fields that confine the fusion plasma.

Tritium Isotope of hydrogen that consisting of one proton and two neutron. Together with deuterium it forms the fuel combination for current fusion devices thanks to their large fusion cross section.

Volume ignition   In the early days of fusion research it was thought that the whole of the fuel had to be compressed to fusion conditions at the end of the compression phase. This concept is called volume ignition.

Waves   Plasmas contain a variety of waves — acoustic, electron plasma — understanding the interplay of all these types of waves is essential for ICF.

X-ray conversion   In indirect-drive schemes the driver does not hit the capsule directly but deposits its energy at the inside of the hohlraum walls. Here the driver energy is converted into x-ray radiation, which then drives the ICF process.

Appendix A

Appendix

A.1 Predicted Energy Consumption and Resources

Fusion research is gaining public awareness as the global energy problem becomes a more and more pressing issue. There are two reasons for that: the Earth's population continues to grow and the consumption of energy per person still increases. As Fig. A.1 shows the per capita energy consumption in the industrialized world is many times that in developing countries.

The Department of Energy (DOE) writes in its "International Energy Outlook 2004" the following:

> The IEO2004 projections indicate continued growth in world's energy use, including large increases for the developing economies

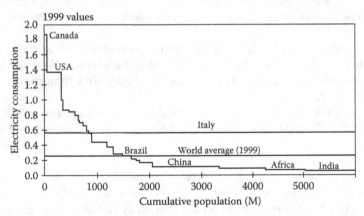

Figure A.1. Electricity consumption per head of population. The units of the electricity consumption are kWh/h/capita. Source: Energy Information Administration, US Department of Energy, 2001.

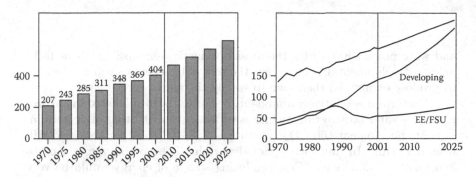

Figure A.2. World energy consumption by regions 1970–2025. Source: Energy Information Administration, US Department of Energy, 2004

of Asia. ... In the developing world as a whole, primary energy consumption is projected to grow at an average annual rate of 2.7 percent between 2001 and 2025 ... in the industrialized world ... energy use expected to grow at 1.2 percent per year ...

(see Figs. A.2a and A.2b). The DOE continues

Oil is expected to remain the dominant energy fuel.

The problem with this is apart from the limited supply the carbon dioxide emission. The world's carbon dioxide emissions are expected to increase from 23,899 million metric tons in 2001 to 37,124 million metric

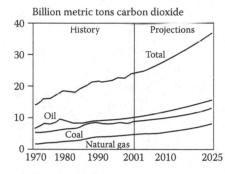
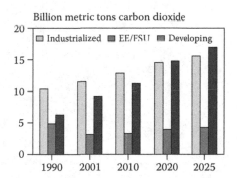

Figure A.3. a) World energy-related carbon dioxide emission by fuel type. b) World energy-related Carbon emission by region. Source: Energy Information Administration, US Department of Energy, 2004.

and wave power. Even with the most optimistic assumptions about technologic development it is clear that the renewable energies will not be able to produce enough on their own to supply the entire energy needs.

Deuterium is relatively abundant, because 1 part in 5000 of the hydrogen in seawater is deuterium, which is equivalent to 10^{15} tons of deuterium.

At the moment the CO_2 emission is 2×10^9 tons per year and still heavily raising. In times when global warming is becoming a major problem for the Earth, a non-CO_2 producing source of energy would be very advantageous. If the right kind of materials are chosen, the radiation hazard presented by fusion power plants can potentially be thousands of times smaller than that of fission plants.

Appendix B

B.1 Constants

Name	Symbol	Value (SI)	Value (cgs)
Boltzmann constant	k_B	1.38×10^{-23} J K^{-1}	1.38×10^{-16} erg K^{-1}
Electron charge	e	1.6×10^{-19} C	4.8×10^{-10} statcoul
Electron mass	m_e	9.1×10^{-31} kg	9.1×10^{-28} g
Proton mass	m_p	1.67×10^{-27} kg	1.67×10^{-24} g
Planck constant	h	6.63×10^{-34} J s	6.63×10^{-27} erg s
Speed of light	c	3×10^8 ms^{-1}	3×10^{10} cm/s
Dielectrical constant	ϵ_0	8.85×10^{-12} F m^{-1}	-
Permeability constant	μ_0	$4\,\pi \times 10^{-7}$	-
Mass ratio	m_p/m_e	1837	1837
Temperature = 1eV	e/k_B	11604	11604
Avogadro number	N_A	6.02×10^{23} mol^{-1}	6.02×10^{23} mol^{-1}
Atmospheric pressure	1 atm	1.013×10^5 Pa	1.013×10^6 dyne/cm^2

Quantity	Symbol		
Debye length	λ_D	$\left(\dfrac{\epsilon_0 k_B T_e}{e^2 n_e}\right)^{\frac{1}{2}}$	$\left(\dfrac{k_B T_e}{4\pi e^2 n_e}\right)^{\frac{1}{2}}$
Number of particles in Debye sphere	N_D	$\dfrac{4\pi}{3}\lambda_D^3 n_e$	$\dfrac{4\pi}{3}\lambda_D^3 n_e$
Electron plasma frequency	ω_p	$\left(\dfrac{e^2 n_e}{\epsilon_0 m_e}\right)^{\frac{1}{2}}$	$\left(\dfrac{4\pi e^2 n_e}{m_e}\right)^{\frac{1}{2}}$
Ion plasma frequency	ω_{pi}	$\left(\dfrac{Z^2 e^2 n_i}{\epsilon_0 m_i}\right)^{\frac{1}{2}}$	$\left(\dfrac{4\pi Z^2 e^2 n_i}{m_i}\right)^{\frac{1}{2}}$
Thermal velocity	$v_{te} = \omega_p \lambda_D$	$\left(\dfrac{k_B T_e}{m_e}\right)^{\frac{1}{2}}$	$\left(\dfrac{k_B T_e}{m_e}\right)^{\frac{1}{2}}$
Electron-ion collision rate	ν_{ei}	$\dfrac{\pi^{\frac{3}{2}} n_e Z e^4 ln\Lambda}{\sqrt{2}(4\pi\epsilon_0)^2 m_e^2 v_{te}^3)}$	$\dfrac{4(2\pi)^{\frac{1}{2}} n_e Z e^4 ln\Lambda}{3 m_e^2 v_{te}^3)}$
Coulomb logarithm	$ln\Lambda$	$ln\dfrac{9N_D}{Z}$	$ln\dfrac{9N_D}{Z}$

EOS	Equation of state
ICF	Inertial confinement fusion
ILE	Institute of Laser Engineering, Osaka University
FWHM	Full width half maximum
FAO	Final optics assemblies
KDP	Potassium dihydrogen phosphate
LULI	Laboratoire d'Utilisation de Laser Intense, France
LMJ	Laser Megajoule
LLE	University of Rochester, Laboratory of Laser Energetics
LLNL	Lawrence Livermore National Laboratory
NIF	National Ignition Facility
PAM	Preamplified modules
RRP	Random phase plates
RAL	Rutherford Appleton Laboratory
RT	Rayleigh-Taylor (instability)
SRS	Stimulated Raman scattering

Wigner extended the formula in 1937. This approach explains the difference between the mass of a nucleus and the mass of its constituent protons and neutrons, the binding energy, as being the result of energies associated the interaction of the nucleons.

The semiempirical mass formula of the nucleus is given by

$$M = N m_n + Z m_p - a_\nu + a_s A^{2/3} + a_c \frac{Z(Z-1)}{A^{1/3}} + a_a \frac{(N-Z)^2}{A} + \frac{a_p \delta}{A^{3/4}},$$

where m_n and m_p are the neutron and the proton mass, and a_ν, a_s, a_c, a_a, and a_p are constants found by fitting experimental binding energies. The best fitting values are

$$a_\nu = 15.282 \text{ MeV}$$
$$a_s = 16.060 \text{ MeV}$$
$$a_c = 0.6876 \text{ MeV}$$
$$a_a = 22.409 \text{ MeV}$$
$$a_p = 16.738 \text{ MeV}.$$

The terms in the equation describe the following:

- The volume term $a\nu$ arises from the interaction of the nucleons through the strong force. The number of interactions is $A(A-1)/2$ so this form of the volume term presumes saturation.
- The surface term $a_s A^{2/3}$ is a correction to the volume term to take into account that the nucleons at the surface of the nucleus do not have the same level of interactions as nucleons in the interior of the nucleus. This term is proportional to the surface area of the nucleus which is proportional to $A^{2/3}$.
- The Coulomb term $a_c \frac{Z(Z-1)}{A^{1/3}}$ represents the energy incorporated in the nucleus as a result of the positive charge. This energy is proportional to the square of the charge and inversely proportional to the radius of the nucleus (i.e. $\sim A^{-1/3}$).
- The asymmetry term $a_a \frac{(N-Z)^2}{A}$ reflects the stability of nuclei with the proton and neutron numbers being approximately equal.
- The odd-even term $\frac{a_p \delta}{A^{3/4}}$ where δ is zero for a nucleus with an odd-even combination of proton and neutron numbers. δ is $+1$ for odd-odd combinations of nucleon numbers and -1 for even-even combinations.

	Los Alamos, USA Sandia, USA	Eulerian	diffusion or transport	
HISHO	Osaka Univ., Japan	Lagrangian	Multigroup diffusion	SESAME
TRITON	Moscow, Russia	Lagrangian	Multigroup diffusion	"Real"
ARWEN	DENIM, Spain	Eulerian or ALE	Multigroup diffusion	QEOS
SARA	DENIM, Spain	Eulerian or ALE		SESAME
MULTI	MPQ, Germany	ALE	Multigroup diffusion	SESAME
FCI	Limeil, France	Lagrangian	Multigroup diffusion	SESAME
DUED	ENEA, Italy	Lagrangian	Multigroup diffusion	SESAME
COBI	ENEA, Italy	Lagrangian	Multigroup diffusion	SESAME
CASTOR	UKAEA, UK	Eulerianian	One group diffusion	Ideal gas

This list shows some of the most used two-dimensional integrated codes in the context of ICF; it is adpated from Velarde et al. (2005). There are many more codes that concentrate on certain aspects in the ICF process.

References

Andre, K., and Betti, R. (2004). *Phys. Plasmas*, **11**, 5.

Andre, M., Babonneau, D., Bayer, C., Bernard, M., Bocher, J. L., Bruneau, J., Coudeville, A., Coutant, J., Dautray, R., Decoster, A., Decroisette, M., D. Desenne, J. M. Dufour, Garconnet, P., Holstein, P. A., Jadaud, J. P., Jolas, A., Juraszek, D., Lachkar, J., Lascaux, P., Lebreton, J. P., Louisjacquet, M., Meyer, B., Mucchielli, F., Rousseaux, C., Schirmann, D., Schurtz, G., Vernon, D., and Watteau, J. P. (1994). *Laser Part. Beams*, **12**, 329.

Andre, M., Cavailler, C., and Jequier, F. (2003). *Le Vide*, **307**, 13.

ARIES (2004). *Fusion Science & Technology*, **46**.

Atzeni, S. (1995). *Jpn.J. Appl. Phys.*, **34**, 1980.

Atzeni, S. (1999). *Phys. Plasmas*, **7**, 3316.

Atzeni, S., Champi, M. L., Piri, A. R., Temporal, M., ter Vehn, J. Meyer, Basko, M. M., Pukhov, A., Rickert, A., Maruhn, J., Kang, K. H., Lutz, K.-J., Ramis, R., Ramirez, R., Sanz, J., and Ibanez, L. F. 1997. *Page 7 of: Fusion Energy 1996: Proceedings of the Sixteenth International Conference Montreal.*

Atzeni, S., Temporal, M., Piriz, A. R., Basko, M. M., Maruhn, J., Lutz, K.-J., Ramis, R., Ramirez, R., Honrubia, J., and ter Vehn, J. Meyer. 1998. *Page 161 of: HIDIF study.*

Azechi, H., Tamari, Y., and Shiraga, H. 2003. *Page 131 of:* Hammel, B. A. (ed), *Inertial Fusion Sciences and Applications.*

Bahcall, J. N., and Waxman, E. (2003). *Physics Letters B*, **556**, 1.

Beynon, G. D., and Constantine, G. (1977). *J. Phys.*, **G3**, 81.

Bodner, S. (1974). *Phys. Rev. Lett.*, **33**, 761.

Bornath, T., Schlanges, M., P. Hilse, P., and Kremp, D. (2001). *Phys. Rev. E*, **64**, 026414.

Brown, M. D., and Moak, C. D. (1972). *Phys. Rev. B*, **6**, 90.

Brueckner, K. A., and Jorna, S. (1974). *Rev. Mod. Phys.*, **46**, 325.

Burnam, A. K., Grens, J., and Lilly, E. M. (1987). *J. Vac. Sci. Technol. A*, **5**, 3417.

Bychenkov, V. Y., Rozmus, W., Tikhonchuk, V. T., and Brantov, A. V. (1995). *Phys. Rev. Lett.*, **75**, 4405.

Cable, M. D. (1995). *Page 191 of:* Hooper, M. B. (ed), *Laser Plasma Interactions : Inertial Confinement Fusion.* IOP, Bristol.

Caruso, A., and Pais, V. A. (1996). *Nucl. Fusion*, **36**, 745.

Caruso, A., Pais, V. A., and Parodi, A. (1992). *Laser Part. Beams*, **10**, 447.

Chabrier, G., Ashcroft, N. W., and Dewitt, H. E. (1992). *Nature*, **360**, 48.

Chaouacha, H. Ben, N. Ben Nessib, N., and S. Sahal-Bréchot, S. (2004). *A&A*, **419**, 771.

Chen, F. F. (1984). *Introduction to Plasma Physics and Controlled Fusion.* Kluwer Academic Pub, Dordrecht.

Cichitelli, L., Eliezer, S., P.Goldsworthy, M., Grenn, F., Hora, H., Ray, P. S., Stening, R. J., and Szichman, H. (1988). *Laser Part. Beams*, **6**, 163.

Cook, R., Overturf, G. E., Buckley, S. R., and McEachern, R. (1994). *J. Vac. Sci. Technol. A*, **9**, 340.

Crawley, R. J. (1986). *J. Vac. Sci. Technol. A*, **3**, 1138.

Dahlberg, J. P., and Gardner, J. H. (1990). *Phys. Rev. A*, **41**, 5695.

Davis, C. C. (1996). *Lasers and Electro-optics Fundamentals and Engineering.* Cambridge University Press, Cambridge.

Dawson, J. M. (1968). *In:* Simon, A., and Thompson, W. (eds), *Advances in plasma physics, vol.1.* Interscience, New York.

Dendy, R. (1994). *Plasma Physics: An Introductionary Course.* Cambridge University Press, Cambridge.

Desselberger, M., and Willi, O. (1993). *Phys. Fluids B*, **5**, 896.

Deutsch, C., Furukawa, H., Mima, K., Murakami, M., and Nishihara, K. (1996). *Phys. Rev. Lett.*, **77**, 2483.

John Wiley & Sons, New York.

Einstein, A. (1917). *Phys. Z.*, **18**, 121.

Eliezer, S. (2002). *The Interaction of High-Power Lasers with Plasmas.* Institute of Physics, London.

Eliezer, S., Ghatak, A., and Ghatak, A. (2002). *Fundamentals of Equations of State.* World Scientific Press, Singapore.

Emery, M. H., Gardner, J. H., and Boris, J. P. (1982). *Phys. Rev. Lett.*, **48**, 677.

Foreman, L. R., Gobby, P., Brooks, P. M., Bush, H., Gomez, V., Elliott, N., Moore, J., Rivera, G., and Salzer, M. (1994). *Fusion Technology.*

Frayley, G. S., Linnebur, E. J., Mason, R. J., and Morse, R. L. (1974). *Phys. Fluids*, **17**, 474.

Gamaly, E. G. (1993). *in Nuclear Fusion by Inertail confinement ed. G.Velarde, Y. Ronen, J.M. Martinez-Val, CRC Press, Boca Raton,* 312.

Gardner, J. H., Bodner, S. E., and Dahlburg, J. P. (1991). *Phys.Fluids B,* **3**, 1070.

Garnier, J. (1999). *Phys. Plasmas,* **6**, 1601.

Gauthier, J. C. (1989). *In:* Hooper, M. B. (ed), *Laser—Plasma Interactions 4, Proceedings of 35th Scottish Universities Summer School in Physics 1988.* Scottish Universities Summer School in Physics Publications, Edinburgh.

Ginzburg, V. L. (1961). *Propagation of Electromagnetic Waves in Plasmas.* Gordon and Breach, New York.

Goldston, R. J., and Rutherford, G. A. (1996). *Introduction to Plasma Physics.* IOP, Bristol.

Goncharov, V. N., Knauer, J. P., McKenty, P. W., Radha, P. B., Sangster, T. C., Skupsky, S., Betti, R., McCrory, R. L., and Meyerhofer, D. D. (2003). *Phys. Plasmas,* **10**, 1906.

Gregori, G., Glenzer, S. H., Knight, J., Niemann, C., Price, D., Froula, D. H., Edwards, M. J., Town, R. P., Brantov, A., Rozmus, W., and Bychenkov, V. Y. (1942). *Phys. Rev. Lett.*, **19**, 302.

Gross, R. A., and Chu, C. K. (1969). Plasma Shock waves. *Adv. Plasma Phys.*, **2**, 139.

Haan, S. W. 2003. *Page 55 of:* Hammel, B. A. (ed), *Inertial Fusion Sciences and Applications*.

Hammel, B. A. (1994). *Phys. Plasmas*, **1**, 1662.

Hammel, B. A., Meyerhofer, D. D., Meyer-ter-Vehn, J., and Azechi, H. (eds) (2004). *Inertial Fusion Sciences and Applications 2003*. American Nuclear Society, Illinois.

Hansen, P., McDonald, I. R., and Vieillefosse, P. (1979). *Phys. Rev. A*, **20**, 2590.

Hatchett, S. P., and Rosen, M. D. 1993. *UCRL-Report JC108348*. Lawrence Livermore.

Hazeltine, R. D., and Meiss, J. D. (2003). *Plasma Confinement*. Dover Pub., Dover.

Henderson, D. B. (1974). *Phys. Rev. Lett.*, **33**.

Hiverly, L. M. (1977). *Nucl. Fusion*, **17**, 873.

Hodgson, P. E., Gadioli, E., and Erba, E. Gadioli (1997). Oxford University Press, Oxford.

Hoffer, J. K. (1992). *In: Proc. 14th Int. Conf. on Plasma Physics and Controlled Nuclear Fusion, Wuerzburg, Germany*. World Scientific, Singapore.

Ichimaru, S. (1982). *Rev. Mod. Phys.*, **54**, 1017.

Kane, J., Arnett, D., Remington, B. A., Glendinning, S. G., Bazan, G., Drake, R. P., and Fryxell, B. A. (2000). *Astrophys. J. Suppl.*, **127**, 365.

Kauffman, R. (1991). *In: Handbook of Plasma Physics Vol 3*. North Holland, Amsterdam.

Kelvin, Lord (1910). *Mathematical and Physical Papers iv, Hydrodynamics and General Dynamics*. Cambridge University Press, Cambridge, England.

Kidder, R. E. (1974). *Nucl. Fusion*, **14**, 953.

Kilkenny, J. K., Glendinning, S. G., Haan, S. W., Hammel, B. A., Lindl, J. D., Munro, D., Remington, B. A., Weber, S. V., Knauer, J. P., and Verdon, C. P. (1994). *Phys. Plasmas*, **1**, 1379.

Kirkpatrick, R. C. (1979). *Nucl. Fusion*, **19**, 69.

Kitagawa, Y. (2002). *Phys. Plasmas*, **9**, 2202.

(2001).   ,   , 993.

Kruer, W. L. (1988). *The Physics of Laser Plasma Interactions*. Addison-Wesley, Redwood City.

Kubo, M., Harada, Y., Kawakatsu, T., and Yonemoto, T. (2001). *J. Chem. Engeneering Japan*, **34**, 1506.

Lawson, J. D. (1957). *Proc. Phys. Soc. London, Sect.B*, **70**, 6.

Lehmberg, R. H. (1987). *J. Appl. Phys.*, **62**, 2680.

Lelevier, R., Lasher, G., and Bjorkland, F. 1955. *Lawrence Livermore Laboratory Report UCRL-4459*.

Lifshitz, E. M., and Pitaevskii, L. P. (1981). *Physical Kinetics*. Pergamon Press, Oxford.

Lindl, J. D. (1995). *Phys. Plasmas*, **2**, 3933.

Lindl, J. D., and McCrory, R. L. (1993). *Il Nouvo Cimento A*, **106**, 1467.

Lindl, J. D., McCrory, R. L., and Campbell, E. M. (1992). *Physics Today*, 32.

Liu, C. S., and Tripathi, V. K. (1995). *Interaction of Electromagnetic Waves with Electron Beams and Plasmas*. World Scientific, Singapore.

Mahdy, A. I., Takabe, H., and Mima, K. (1999). *Nucl. Fusion*, **39**, 467.

Martinez-Val, J. M., Velarde, G., and Ronen, Y. (1993). An introduction to nuclear fusion by inertial confinement. *Pages 1–42 of:* Velarde, G., Ronen, Y., and Martinez-Val, J.M. (eds), *Nuclear Fusion by Inertial Confinement*. CRC Press, Boca Raton.

Mason, R. J., and Morse, R.L. (1975). *Phys. Fluids*, **18**, 814.

Matsui, H., Eguchi, T., Kanabe, T., Yamanaka, M., Nakatsuka, M., Izawa, Y., and Nakai, S. (2000). *Rcv. Laser Eng.*, **28**, 176.

Max, C. E., McKee, C. F., and Mead, W. C. (1980). *Phys.Fluids*, **23**, 1620.

McCrory, R. (2003). *in Inertial Fusion Sciences and Applications ed. B.A.Hammel et al*, 3.

McCrory, R. L., Bahr, R. E., Betti, R., Boehly, T. R., Collins, T. J. B., Craxton, R. S., Delettrez, J. A., Donaldson, W. R., Epstein, R., Frenje, J., Glebev, V. Y., Goncharov, V. N., Gotchev, O. G., Gram, R. Q., Harding, D. R., Hicks, D. G., Jaanimagi, P. A., Keck, R. L., kelly, J. H., Knauer, J. P., Li, C. K., Loucks, S. J., Lund, L. D., Marshall, F. J., Kenty, P. W. Mc, Meyerhofer, D. D., Morse, S. F. B., Petrasso, R. D.,

McKenty, P. W., Goncharov, V. N., P.Town, R. J., Skupsky, S., Betti, R., and McCrory, R. L. (2001). *Phys. Plasmas*, **8**, 2315.

McQuillan, B. W., and Takagi, M. (2002). *Fusion Sci. Techn.*, **41**, 209.

Meyer-ter-Vehn, J. (1982). *Nucl. Fusion*, **22**, 561.

Moir, R. W. (1994). *Fusion Technology*, **25**, 5.

More, R. M., Zinamon, Z., Warren, K. H., Falcone, R., and Murnane, M. (1988). *J. de Physique*, **49**, C7.

Musinski, D. L., Henderson, T. M., Simms, R. J., Pattinson, T. R., and Jacobs, R. B. (1980). *J. Appl. Phys.*, **51**, 1394.

Nakai, S. 1994. *In: Proc. of 15th Int. Conf. Plasma Physics and Controlled Nuclear Fusion Research.*

Nakai, S., and Mima, K. (2004). *Rep. Prog. Phys.*, **67**, 321.

Nikroo, A., Pontelandolfo, J. M., and Castillo, E. R. (2002). *Fusion Sci. Tech.*, **41**, 220.

Nishikawa, K. (1968). *J. Phys. Soc. Jap.*, **24**, 1153.

Norimatsu, T., Chen, C. M., Nakajima, K., Takagi, M., Izawa, Y., Yamanaka, T., and Nakai, S. (1994). *J. Vac. Sci. Technol. A*, **12**, 1293.

Nuckolls, J., Wood, L., Thiessen, A., and Zimmerman, G. (1972). *Nature*, **239**, 139.

Nuckolls, J. H. (1994). Plenum Press, New York.

Obenschein, S. P. (1986). *Phys. Rev. Lett.*, **56**, 2807.

Olson, R. E., Leeper, R. J., Nobile, A., and Oertel, J. A. (2003). *Phys. Rev. Lett.*, **91**, 235002.

Perry, M. D., and Mourou, G. (1994). *Science*, **246**, 917.

Peter, T., and ter Vehn, J. Meyer (1991). *Phys. Rev. A*, **43**, 2015.

Petzoldt, R. W., Goodin, D. T., Nikroo, A., Stephens, E., Sigel, N., Alexander, N. B., Raffray, A. R., Mau, T. K., Tillack, M., Najmabadi, F., Krasheninnikov, S. I., and Gallix, R. (2002). *Nucl. Fusion*, **42**, 1351.

Pfalzner, S., and Gibbon, P. (1996). *Many-Body Tree Methods Physics.* Cambridge University Press, Cambridge.

Pfalzner, S., and Gibbon, P. (1998). *Phys. Rev. E*, **57**, 4698.

Piriz, A. R., and Atzeni, S. (1994). *Plasma Phys. Controll. Fusion*, **36**, 451.

Regan, S. P., Marozas, J. A., Craxton, R. S., Kelly, J. H., Donaldson, W. R., Jaanimagi, P. A., Jacobs-Perkins, D., Keck, R. L., Kessler, T. J., Meyerhofer, D. D., Sangster, T. C., Seka, W., Smalyuk, V. A., Skupsky, S., and Zuegel, J. D. (2005). *J. Opt. Soc. Am B*, **22**.

Remington, B. A., Weber, S. V., Haan, S. W., Kilkenny, J. D., Glendinning, Wallace, R. J., Goldstein, W. H., Wilson, B. G., and Nash, J. K. (1993). *Phys. Fluids B*, **5**, 2589.

Richtmyer, R. D. (1960). *Commun. Pure Appl. Math.*, **13**, 297.

Rickman, W. S., and Goodin, D. T. (2003). *Fusion Sci. Techn.*, **43**, 353.

Rose, H. A., and Dubois, F. D. (1994). *Phys. Rev. Lett*, **72**, 2883–2886.

Rose, S. (1988). SUSSP Publications, Edinburgh.

Rosen, M. D., and Lindl, J. D. 1983. *Laser Program Annual Report 83 UCRL-50021-83*.

Roth, M., Cowan, T. E., Key, M. H., Hatchett, S. P., Brown, C., Fountain, W., Johnson, J., Pennington, D. M., Snavely, R. A., Wilks, S. C., Yasuike, K., Ruhl, H., Pegoraro, F., Bulanov, S. V., Campbell, E. M., Perry, M. D., Powell, H., Rosen, M. D., and Lindl, J. D. (2001). *Phys. Rev. Lett.*, **86**, 436.

Rothenberg, J. E. (1997). *J. Opt. Soc. Am. B*, **14**, 1664.

Sanz, J. (1994). *Phys. Rev. Lett*, **73**, 27007.

Sethian, J. D., Friedman, M., Giuliani, J. L., Lehmberg, R. H., Obenschain, S. P., Kepple, P., Wolford, M., Hegeler, F., Swanekamp, S. B., Weidenheimer, D., Welch, D., Rose, D. V., and Searles, S. (2003). *Phys. Plasmas*, **10**, 2142.

Sid, A. (2003). *Phys. Plasmas*, **10**, 214.

Singh, S. (1987). *Handbook of Laser Science and Technology, Vol.3*. CRC Press, Boca Raton, FL.

Skupsky, S. (2004). *Phys. Plasmas*, **11**, 2763.

Skupsky, S., and Craxton, R. S. (1999). *Phys. Plasmas*, **6**, 2157.

Soures, J. M., McCrory, R. L., Verdon, C. P., Babushkin, A., Bahr, R. E., Boehly, T. R., Boni, R., Bradley, D. K., Brown, D. L., Craxton, R. S., Delettrez, J. A., Donaldson, W. R., Jaanimagi, R. Epsteinand P. A., and Jacobs, S. (1996). *Phys. Plasmas*, **3**, 2108.

Spitzer, L., and Harm, R. (1953). *Phys. Rev.*, **89**, 977.

Tabak, M., Munro, D. H., and Lindl, J. D. (1990). *Phys. Fluids B*, **2**, 1007.

Tabak, M., Hammer, J., Glinsky, M. E., Kruer, W. L., Wilks, S. C., Woodworth, J., Campbell, E. M., Perry, M. D., and Mason, R. J. (1994). *Phys. Plasmas*, **1**, 1626.

Takabe, H., Mimo, K., Montierth, L., and Morse, R. L. (1985). *Phys. Fluids*, **28**, 3676.

Tanaka, K. A., Kodama, R., Fujita, H., Heya, M., Izumi, N., Kato, Y., Kitagawa, Y., Mima, K., Miyanaga, N., Norimatsu, T., Pukhov, A., Sunahara, A., Takahashi, K., Allen, M., and al, H. Habaraet (2000). *Phys. Plasmas*, **7**, 2014.

Taylor, G. (1950). *Proc. Roy. Soc. A*, **201**, 192.

Town, R. P. J., and Bell, A. R. (1991). *Phys. Rev. Lett.*, **67**, 1863.

Tsubakimoto, K., Jitsuno, T., Miyanaga, N., Nakatsuda, M., Kanabe, T., and Nakai, S. (1993). *Opt. Commun.*, **103**, 185.

Velarde, G., Martinez-Val, J. M., and Eliezer, S. 2005. *Prospects on the use of inertial nuclear fusion.*

Watt, R. G., Cobble, J., and Dubois, D. F. (1996). *Phys. Plasmas*, **3**, 1091.

Welch, D. R., Rose, D. V., Oliver, B. V., and Clark, R. E. (2001). *Nucl. Inst. Meth. Phys. Res. A*, **242**, 134.

Widner, M. M. 1979. *Sandia-Report SAND79-2454.*

Woodworth, J. G., and Meier, W. R. (1997). *Fusion Techn.*, **31**, 280.

Y. Lin, T.J. Kessler, G.N. Lawrence (1996). *Opt.Lett*, **21**, 1703.

Yabe, T. (1993). *Pages 269–292 of:* Velarde, G., Ronen, Y., and Martinez-Val, J.M. (eds), *Nuclear Fusion by Inertial Confinement.* CRC Press, Boca Raton.

Yamanaka, C. (1989a). *in Proc. 5th Int. Conf. Emerging Nuclear Energy Systems, ed. by U.V. Mollendorf and B. Goeld, World Scientific, Singapore*, 125.

Yamanaka, C. (1989b). *Page 3 of:* et al., G. Velarde (ed), *Laser Interaction with Matter.* World Scientific, Singapore.

Yamanaka, T. (1989c). *Page 105 of:* et al., G. Velarde (ed), *Laser Interaction with Matter.* World Scientific, Singapore.

Yu, S., Anders, A., Bieniosek, F. M., Eylon, S., Henestroza, E., Roy, P., Shuman, D., Waldron, W. L., Houck, T., Sharp, W., Rose, D., Dale, W.,

Faltens, A., Friedman, A., Grote, D. P., Heitzenroeder, P., Henestroza, E., Kaganovich, I., Kwan, J. W., Latkowski, J. F., Lee, E. P., Logan, B. G., Peterson, P. F., Rose, D., Roy, P.K., Sabbi, G.-L., Seidl, P.-A.; Sharp, W. M., and Welch, D. R. (2005). *Nucl. Instr. Meth.*, **544**, 294.

Printed in the United States
by Baker & Taylor Publisher Services